Community Detection and Mining in Social Media

Community Detection and Mining in Social Media

Synthesis Lectures on Data Mining and Knowledge Discovery

Editor
Jiawei Han, *University of Illinois at Urbana-Champaign*
Lise Getoor, *University of Maryland*
Wei Wang, *University of North Carolina, Chapel Hill*
Johannes Gehrke, *Cornell University*
Robert Grossman, *University of Illinois, Chicago*

Synthesis Lectures on Data Mining and Knowledge Discovery is edited by Jiawei Han, Lise Getoor, Wei Wang, and Johannes Gehrke. The series publishes 50- to 150-page publications on topics pertaining to data mining, web mining, text mining, and knowledge discovery, including tutorials and case studies. The scope will largely follow the purview of premier computer science conferences, such as KDD. Potential topics include, but not limited to, data mining algorithms, innovative data mining applications, data mining systems, mining text, web and semi-structured data, high performance and parallel/distributed data mining, data mining standards, data mining and knowledge discovery framework and process, data mining foundations, mining data streams and sensor data, mining multi-media data, mining social networks and graph data, mining spatial and temporal data, pre-processing and post-processing in data mining, robust and scalable statistical methods, security, privacy, and adversarial data mining, visual data mining, visual analytics, and data visualization.

Community Detection and Mining in Social Media

Lei Tang and Huan Liu

ISBN: 978-3-031-00772-9   paperback
ISBN: 978-3-031-01900-5   ebook

DOI 10.1007/978-3-031-01900-5

A Publication in the Springer series
SYNTHESIS LECTURES ON DATA MINING AND KNOWLEDGE DISCOVERY

Lecture #3
Series Editor: Jiawei Han, *University of Illinois at Urbana-Champaign*
          Lise Getoor, *University of Maryland*
          Wei Wang, *University of North Carolina, Chapel Hill*
          Johannes Gehrke, *Cornell University*
          Robert Grossman, *University of Illinois, Chicago*
Series ISSN
Synthesis Lectures on Data Mining and Knowledge Discovery
Print 2151-0067   Electronic 2151-0075

Synthesis Lectures on Data Mining and Knowledge Discovery

Editor
Jiawei Han, *University of Illinois at Urbana-Champaign*
Lise Getoor, *University of Maryland*
Wei Wang, *University of North Carolina, Chapel Hill*
Johannes Gehrke, *Cornell University*
Robert Grossman, *University of Illinois, Chicago*

Synthesis Lectures on Data Mining and Knowledge Discovery is edited by Jiawei Han, Lise Getoor, Wei Wang, and Johannes Gehrke. The series publishes 50- to 150-page publications on topics pertaining to data mining, web mining, text mining, and knowledge discovery, including tutorials and case studies. The scope will largely follow the purview of premier computer science conferences, such as KDD. Potential topics include, but not limited to, data mining algorithms, innovative data mining applications, data mining systems, mining text, web and semi-structured data, high performance and parallel/distributed data mining, data mining standards, data mining and knowledge discovery framework and process, data mining foundations, mining data streams and sensor data, mining multi-media data, mining social networks and graph data, mining spatial and temporal data, pre-processing and post-processing in data mining, robust and scalable statistical methods, security, privacy, and adversarial data mining, visual data mining, visual analytics, and data visualization.

Community Detection and Mining in Social Media
Lei Tang and Huan Liu

ISBN: 978-3-031-00772-9   paperback
ISBN: 978-3-031-01900-5   ebook

DOI 10.1007/978-3-031-01900-5

A Publication in the Springer series
SYNTHESIS LECTURES ON DATA MINING AND KNOWLEDGE DISCOVERY

Lecture #3
Series Editor: Jiawei Han, *University of Illinois at Urbana-Champaign*
        Lise Getoor, *University of Maryland*
        Wei Wang, *University of North Carolina, Chapel Hill*
        Johannes Gehrke, *Cornell University*
        Robert Grossman, *University of Illinois, Chicago*
Series ISSN
Synthesis Lectures on Data Mining and Knowledge Discovery
Print 2151-0067   Electronic 2151-0075

Community Detection and Mining in Social Media

Lei Tang
Yahoo! Labs

Huan Liu
Arizona State University

SYNTHESIS LECTURES ON DATA MINING AND KNOWLEDGE DISCOVERY
#3

ABSTRACT

The past decade has witnessed the emergence of participatory Web and social media, bringing people together in many creative ways. Millions of users are playing, tagging, working, and socializing online, demonstrating new forms of collaboration, communication, and intelligence that were hardly imaginable just a short time ago. Social media also helps reshape business models, sway opinions and emotions, and opens up numerous possibilities to study human interaction and collective behavior in an unparalleled scale. This lecture, from a data mining perspective, introduces characteristics of social media, reviews representative tasks of computing with social media, and illustrates associated challenges. It introduces basic concepts, presents state-of-the-art algorithms with easy-to-understand examples, and recommends effective evaluation methods. In particular, we discuss graph-based community detection techniques and many important extensions that handle dynamic, heterogeneous networks in social media. We also demonstrate how discovered patterns of communities can be used for social media mining. The concepts, algorithms, and methods presented in this lecture can help harness the power of social media and support building socially-intelligent systems. This book is an accessible introduction to the study of *community detection and mining in social media*. It is an essential reading for students, researchers, and practitioners in disciplines and applications where social media is a key source of data that piques our curiosity to understand, manage, innovate, and excel.

This book is supported by additional materials, including lecture slides, the complete set of figures, key references, some toy data sets used in the book, and the source code of representative algorithms. The readers are encouraged to visit the book website for the latest information:

http://dmml.asu.edu/cdm/

KEYWORDS

social media, community detection, social media mining, centrality analysis, strength of ties, influence modeling, information diffusion, influence maximization, correlation, homophily, influence, community evaluation, heterogeneous networks, multi-dimensional networks, multi-mode networks, community evolution, collective classification, social dimension, behavioral study

To my parents and wife — LT

To my parents, wife, and sons — HL

Contents

Acknowledgments

It is a pleasure to acknowledge many colleagues who made substantial contributions in various ways to this time-consuming book project. The members of the Social Computing Group, Data Mining and Machine Learning Lab at Arizona State University made this project enjoyable. They include Ali Abbasi, Geoffrey Barbier, William Cole, Gabriel Fung, Huiji Gao, Shamanth Kumar, Xufei Wang, and Reza Zafarani. Particular thanks go to Reza Zafarani and Gabriel Fung who read the earlier drafts of the manuscript and provided helpful comments to improve the readability.

We are grateful to Professor Sun-Ki Chai, Professor Michael Hechter, Dr. John Salerno and Dr. Jianping Zhang for many inspiring discussions. This work is part of the projects sponsored by grants from AFOSR and ONR.

We thank Morgan & Claypool and particularly executive editor Diane D. Cerra for her help and patience throughout this project. We also thank Professor Nitin Agarwal at University of Arkansas at Little Rock for his prompt answers to questions concerning the book editing.

Last and foremost, we thank our families for supporting us through this fun project. We dedicate this book to them, with love.

Lei Tang and Huan Liu
August, 2010

CHAPTER 1

Social Media and Social Computing

1.1 SOCIAL MEDIA

The past decade has witnessed a rapid development and change of the Web and the Internet. Numerous participatory web applications and social networking sites have been cropping up, drawing people together and empowering them with new forms of collaboration and communication. Prodigious numbers of online volunteers collaboratively write encyclopedia articles of previously unthinkable scopes and scales; online marketplaces recommend products by tapping on crowd wisdom via user shopping and reviewing interactions; and political movements benefit from new forms of engagement and collective actions.

Table 1.1 lists various social media, including blogs, forums, media sharing platforms, microblogging, social networking, social news, social bookmarking, and wikis. Underneath their seemingly different facade lies a common feature that distinguishes them from the classical web and traditional media: *consumers of content, information and knowledge are also producers.*

Table 1.1: Various Forms of Social Media	
Blog	Wordpress, Blogspot, LiveJournal, BlogCatalog
Forum	Yahoo! answers, Epinions
Media Sharing	Flickr, YouTube, Justin.tv, Ustream, Scribd
Microblogging	Twitter, foursquare, Google buzz
Social Networking	Facebook, MySpace, LinkedIn, Orkut, PatientsLikeMe
Social News	Digg, Reddit
Social Bookmarking	Del.icio.us, StumbleUpon, Diigo
Wikis	Wikipedia, Scholarpedia, ganfyd, AskDrWiki

In traditional media such as TV, radio, movies, and newspapers, it is only a small number of "authorities" or "experts" who decide which information should be produced and how it is distributed. The majority of users are consumers who are separated from the production process. The communication pattern in the traditional media is one-way traffic, from a centralized producer to widespread consumers.

A user of social media, however, can be both a consumer and a producer. With hundreds of millions of users active on various social media sites, everybody can be a media outlet (Shirky,

Table 1.2: Top 20 Websites in the US			
Rank	Site	Rank	Site
1	google.com	11	blogger.com
2	facebook.com	12	msn.com
3	yahoo.com	13	myspace.com
4	youtube.com	14	go.com
5	amazon.com	15	bing.com
6	wikipedia.org	16	aol.com
7	craigslist.org	17	linkedin.com
8	twitter.com	18	cnn.com
9	ebay.com	19	espn.go.com
10	live.com	20	wordpress.com

2008). This new type of mass publication enables the production of timely news and grassroots information and leads to mountains of *user-generated contents*, forming *the wisdom of crowds*. One example is the London terrorist attack in 2005 (Thelwall, 2006) in which some witnesses blogged their experience to provide first-hand reports of the event. Another example is the bloody clash ensuing the Iranian presidential election in 2009 where many provided live updates on Twitter, a microblogging platform. Social media also allows collaborative writing to produce high-quality bodies of work otherwise impossible. For example, "since its creation in 2001, Wikipedia has grown rapidly into one of the largest reference web sites, attracting around 65 million visitors monthly as of 2009. There are more than 85,000 active contributors working on more than 14,000,000 articles in more than 260 languages[1]."

Another distinctive characteristic of social media is its rich user interaction. The success of social media relies on the participation of users. More user interaction encourages more user participation, and vice versa. For example, Facebook claims to have more than 500 million active users[2] as of August, 2010. The user participation is a key element to the success of social media, and it has helped push eight social media sites to be among the top 20 websites as shown in Table 1.2 (Internet traffic by Alexa on August 3, 2010). Users are connected though their interactions from which *networks* of users emerge. Novel opportunities arise for us to study human interaction and collective behavior on an unprecedented scale and many computational challenges ensue, urging the development of advanced computational techniques and algorithms.

This lecture presents basic concepts of social network analysis related to community detection and uses simple examples to illustrate state-of-the-art algorithms for using and analyzing social media data. It is a self-contained book and covers important topics regarding community detection in social media. We start with concepts and definitions used throughout this book.

[1] http://en.wikipedia.org/wiki/Wikipedia:About
[2] http://www.facebook.com/press/info.php?statistics

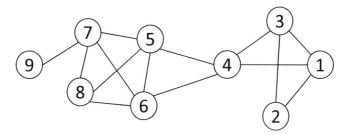

Figure 1.1: A social network of 9 actors and 14 connections. The diameter of the network is 5. The clustering coefficients of nodes 1-9 are: $C_1 = 2/3$, $C_2 = 1$, $C_3 = 2/3$, $C_4 = 1/3$, $C_5 = 2/3$, $C_6 = 2/3$, $C_7 = 1/2$, $C_8 = 1$, $C_9 = 0$. The average clustering coefficient is 0.61 while the expected clustering coefficient of a random graph with 9 nodes and 14 edges is $14/(9 \times 8/2) = 0.19$.

Table 1.3: Adjacency Matrix									
Node	1	2	3	4	5	6	7	8	9
1	-	1	1	1	0	0	0	0	0
2	1	-	1	0	0	0	0	0	0
3	1	1	-	1	0	0	0	0	0
4	1	0	1	-	1	1	0	0	0
5	0	0	0	1	-	1	1	1	0
6	0	0	0	1	1	-	1	1	0
7	0	0	0	0	1	1	-	1	1
8	0	0	0	0	1	1	1	-	0
9	0	0	0	0	0	0	1	0	-

1.2   CONCEPTS AND DEFINITIONS

Network data are different from attribute-value data and have their unique properties.

1.2.1   NETWORKS AND REPRESENTATIONS

A social network is a social structure made of nodes (individuals or organizations) and edges that connect nodes in various relationships like friendship, kinship, etc. There are two common ways to represent a network. One is the graphical representation that is very convenient for visualization. Figure 1.1 shows one toy example of a social network of 9 actors. A social network can also be represented as a matrix (called the *sociomatrix* (Wasserman and Faust, 1994), or *adjacency matrix*) as shown in Table 1.3. Note that a social network is often very sparse, as indicated by many zeros in the table. This sparsity can be leveraged for performing efficient analysis of networks. In the adjacency matrix, we do not specify the diagonal entries. By definition, diagonal entries represent self-links, i.e., a connection from one node to itself. Typically, diagonal entries are set to zero for network

Table 1.4: Nomenclature

Symbol	Meaning
A	the adjacency matrix of a network
V	the set of nodes in the network
E	the set of edges in the network
n	the number of nodes ($n = |V|$)
m	the number of edges ($n = |E|$)
v_i	a node v_i
$e(v_i, v_j)$	an edge between nodes v_i and v_j
A_{ij}	1 if an edge exists between nodes v_i and v_j; 0 otherwise
N_i	the neighborhood (i.e., neighboring nodes) of node v_i
d_i	the degree of node v_i ($d_i = |N_i|$)
geodesic	a shortest path between two nodes
geodesic distance	the length of the shortest path
$g(v_i, v_j)$	the geodesic distance between nodes v_i and v_j

analysis. However, in certain cases, diagonal entries should be fixed to one. Thereafter, unless we specify explicitly, the diagonal entries are zero by default.

A network can be weighted, signed and directed. In a weighted network, edges are associated with numerical values. In a signed network, some edges are associated with positive relationships, some others might be negative. Directed networks have directions associated with edges. In our example in Figure 1.1, the network is undirected. Correspondingly, the adjacency matrix is symmetric. However, in some social media sites, interactions are directional. In Twitter, for example, one user x follows another user y, but user y does not necessarily follow user x. In this case, the follower-followee network is directed and asymmetrical. This lecture will focus on, unless specified explicitly, a simplest form of network, i.e., undirected networks with boolean edge weights, just like the example in Table 1.3. Many of the techniques discussed in this lecture can be extended to handle weighted, signed and directed networks as well.

Notations We use V and E to denote the sets of nodes and edges in a network, respectively. The number of nodes in the network is n, and the number of edges is m. Matrix $A \in \{0, 1\}^{n \times n}$ represents the adjacency matrix of a network. An entry $A_{ij} \in \{0, 1\}$ denotes whether there is a link between nodes v_i and v_j. The edge between two nodes v_i and v_j is denoted as $e(v_i, v_j)$. Two nodes v_i and v_j are *adjacent* if $A_{ij} = 1$. N_i represents all the nodes that are adjacent to node v_i, i.e., the neighborhood of node v_i. The number of nodes adjacent to a node v_i is called its *degree* (denoted as d_i). E.g., in the network in Figure 1.1, $d_1 = 3, d_4 = 4$. One edge is *adjacent* to a node if the node is a terminal node of the edge. E.g., edge $e(1, 4)$ is adjacent to both nodes 1 and 4.

A shortest path between two nodes (say, v_i and v_j) is called a *geodesic*. The number of hops in the geodesic is the *geodesic distance* between the two nodes (denoted as $g(v_i, v_j)$). In the example,

$g(2, 8) = 4$ as there is a geodesic (2-3-4-6-8). The notations that will be used frequently throughout this lecture are summarized in Table 1.4.

1.2.2   PROPERTIES OF LARGE-SCALE NETWORKS

Networks in social media are often very huge, with millions of actors and connections. These large-scale networks share some common patterns that are seldom noticeable in small networks. Among them, the most well-known are: *scale-free distributions*, *the small-world effect*, and *strong community structure*. Networks with non-trivial topological features are called *complex networks* to differentiate them from simple networks such as a lattice graph or random graphs.

Node degrees in a large-scale network often follow a power law distribution. Let's look at Figure 1.2 about how many nodes having a particular node degree for two networks of YouTube[3] and Flickr[4]. As seen in the figure, the majority of nodes have a low degree, while few have an extremely high degree (say, degree > 10^4). In a log-log scale, both networks demonstrate a similar pattern (approximately linear, or a straight line) in node degree distributions (node degree vs. number of nodes). Such a pattern is called a *power law distribution*, or a *scale-free distribution* because the shape of the distribution does not change with scale. What is interesting is that if we zoom into the tail (say, examine those nodes with degree > 100), we will still see a power law distribution. This self-similarity is independent of scales. Networks with a power law distribution for node degrees are called *scale-free networks*.

Another prominent characteristic of social networks is the so-called *small-world effect*. Travers and Milgram (1969) conducted an experiment to examine the average path length for social networks of people in the United States. In the experiment, the subjects chosen were asked to send a chain letter to his acquaintances starting from an individual in Omaha, Nebraska or Wichita, Kansas (then remote places) to the target individual in Boston, Massachusetts. At the end, 64 letters arrived and the average path length fell around 5.5, or roughly 6, hence, the famous "*six degrees of separation*". This result is also confirmed recently in a planetary-scale instant messaging network of more than 180 million people, in which the average path length of any two people is 6.6 (Leskovec and Horvitz, 2008). In order to calibrate this small world effect, some additional measures are defined. The length of the longest geodesic in the network is its *diameter* (Wasserman and Faust, 1994). The diameter of the network in Figure 1.1 is 5 (corresponding to the geodesic distance between nodes 2 and 9). Most real-world large-scale networks observe a small diameter.

Social networks also exhibit a strong *community structure*. That is, people in a group tend to interact with each other more than with those outside the group. Since friends of a friend are likely to be friends as well, this transitivity can be measured by *clustering coefficient*, the number of connections between one's friends over the total number of possible connections among them. Suppose a node

---

[3]http://www.youtube.com/
[4]http://www.flickr.com/

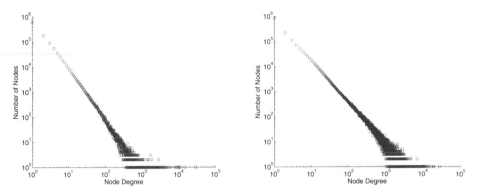

Figure 1.2: Node distributions of a YouTube network with 1, 138, 499 nodes and a Flickr network with 1, 715, 255 edges (based on data from (Mislove et al., 2007)). Both X-axis and Y-axis are on a log scale. For power law distributions, the scatter plot of node degrees is approximately a straight line. The average clustering coefficients are 0.08 and 0.18, whereas if the connections are uniformly random, the expected coefficient in a random graph is 4.6×10^{-6} and 1.0×10^{-5}.

v_i has d_i neighbors, and there are k_i edges among these neighbors, then the clustering coefficient is

$$C_i = \begin{cases} \frac{k_i}{d_i \times (d_i-1)/2} & d_i > 1 \\ 0 & d_i = 0 \ or \ 1 \end{cases} \tag{1.1}$$

Clustering coefficient measures the density of connections among one's friends. A network with communities tends to have a much higher average clustering coefficient than a random network. For instance, in Figure 1.1, node 6 has four neighbors 4, 5, 7 and 8. Among them, we have four connections $e(4, 5), e(5, 7), e(5, 8), e(7, 8)$. Hence, the clustering coefficient of node 6 is $4/(4 \times 3/2) = 2/3$. The average clustering coefficient of the network is 0.61. However, for a random graph with the same numbers of nodes and connections, the expected clustering coefficient is $14/(9 \times 8/2) = 0.19$.

1.3   CHALLENGES

Millions of users are playing, working, and socializing online. This flood of data allows for an unprecedented large-scale social network analysis – millions of actors or even more in one network. Examples include email communication networks (Diesner et al., 2005), instant messaging networks (Leskovec and Horvitz, 2008), mobile call networks (Nanavati et al., 2006), friendship networks (Mislove et al., 2007). Other forms of complex networks, like coauthorship or citation networks, biological networks, metabolic pathways, genetic regulatory networks and food web, are also examined and demonstrate similar patterns (Newman et al., 2006). Social media enables a new laboratory to study human relations.

These large-scale networks combined with unique characteristics of social media present novel challenges for mining social media. Some examples are given below:

- Scalability. The network presented in social media can be huge, often in a scale of millions of actors and hundreds of millions of connections, while traditional social network analysis normally deals with hundreds of subjects or fewer. Existing network analysis techniques might fail when applied directly to networks of this astronomical size.

- Heterogeneity. In reality, multiple relationships can exist between individuals. Two persons can be friends and colleagues at the same time. Thus, a variety of interactions exist between the same set of actors in a network. Multiple types of entities can also be involved in one network. For many social bookmarking and media sharing sites, users, tags and content are intertwined with each other, leading to heterogeneous entities in one network. Analysis of these heterogeneous networks involving heterogeneous entities or interactions requires new theories and tools.

- Evolution. Social media emphasizes timeliness. For example, in content sharing sites and blogosphere, people quickly lose their interest in most shared contents and blog posts. This differs from classical web mining. New users join in, new connections establish between existing members, and senior users become dormant or simply leave. How can we capture the dynamics of individuals in networks? Can we find the die-hard members that are the backbone of communities? Can they determine the rise and fall of their communities?

- Collective Intelligence. In social media, people tend to share their connections. The wisdom of crowds, in forms of tags, comments, reviews, and ratings, is often accessible. The meta information, in conjunction with user interactions, might be useful for many applications. It remains a challenge to effectively employ social connectivity information and collective intelligence to build social computing applications.

- Evaluation. A research barrier concerning mining social media is evaluation. In traditional data mining, we are so used to the training-testing model of evaluation. It differs in social media. Since many social media sites are required to protect user privacy information, limited benchmark data is available. Another frequently encountered problem is the lack of ground truth for many social computing tasks, which further hinders some comparative study of different works. Without ground truth, how can we conduct fair comparison and evaluation?

1.4   SOCIAL COMPUTING TASKS

Associated with the challenges are some pertinent research tasks. We illustrate them with examples.

1.4.1   NETWORK MODELING

Since the seminal work by Watts and Strogatz (1998), and Barabási and Albert (1999), network modeling has gained some significant momentum (Chakrabarti and Faloutsos, 2006). Researchers have observed that large-scale networks across different domains follow similar patterns, such as scale-free distributions, the small-world effect and strong community structures as we discussed in Section 1.2.2. Given these patterns, it is intriguing to model network dynamics of repeated patterns with some simple mechanics. Examples include the *Watts-and-Strogatz model* (Watts and Strogatz, 1998) to explain the small-world effect and the *preferential attachment process* (Barabási and Albert, 1999) that explains power-law distributions. Network modeling (Chakrabarti and Faloutsos, 2006) offers an in-depth understanding of network dynamics that is independent of network domains. A network model can be used for simulation study of various network properties, e.g., robustness of a network under attacks, information diffusion within a given network structure, etc.

When networks scale to over millions and more nodes, it becomes a challenge to compute some network statistics such as the diameter and average clustering coefficient. One way to approach the problem is sampling (Leskovec and Faloutsos, 2006). It provides an approximate estimate of different statistics by investigating a small portion of the original huge network. Others explore I/O efficient computation (Becchetti et al., 2008; Desikan and Srivastava, 2008). Recently, techniques of harnessing the power of distributed computing (e.g., MapReduce mechanism in Hadoop platform) are attracting increasing attention.

1.4.2   CENTRALITY ANALYSIS AND INFLUENCE MODELING

Centrality analysis is about identifying the most "important" nodes in a network (Wasserman and Faust, 1994). Traditional social network analysis hinges upon link structures to identify nodes with high centrality. Commonly used criteria include degree centrality, betweenness centrality, closeness centrality, and eigenvector centrality (equivalent to Pagerank scores (Page et al., 1999)). In social media, additional information is available, such as comments and tags. This change opens up opportunities to fuse various sources of information to study centrality (Agarwal et al., 2008).

A related task is *influence modeling* that aims to understand the process of influence or information diffusion. Researchers study how information is propagated (Kempe et al., 2003) and how to find a subset of nodes that maximize influence in a population. Its sibling tasks include blocking rumor diffusion or detecting cascading behavior by placing a limited number of sensors online (Leskovec et al., 2007b). In the marketing domain, it is also known as *viral marketing* (Richardson and Domingos, 2002) or *word-of-mouth marketing*. It aims to identify influential customers for marketing such that they can effectively influence their friends to achieve the maximum return.

1.4.3   COMMUNITY DETECTION

Communities are also called *groups*, *clusters*, *cohesive subgroups*, or *modules* in different contexts. It is one of the fundamental tasks in social network analysis. Actually, "the founders of sociology claimed that the causes of social phenomena were to be found by studying groups rather than individuals" (Hechter (1988), Chapter 2, Page 15). Finding a community in a social network is to identify a set of nodes such that they interact with each other more frequently than with those nodes outside the group. For instance, Figure 1.3 shows the effect after the nodes are grouped into two different sets

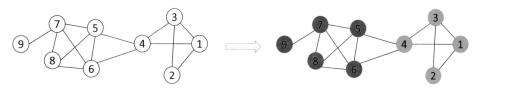

Figure   1.3: Grouping   Effect.   Two   communities   are   identified   via   modularity   optimization (Newman and Girvan, 2004): {1, 2, 3, 4}, and {5, 6, 7, 8, 9}.

based on modularity optimization (discussed later in Section 3.3.5), with each group in a different
color.

     Community detection can facilitate other social computing tasks and is applied in many real-
world applications. For instance, the grouping of customers with similar interests in social media
renders efficient recommendations that expose customers to a wide range of relevant items to enhance
transaction success rates. Communities can also be used to compress a huge network, resulting in
a smaller network. In other words, problem solving is accomplished at group level, instead of node
level. In the same spirit, a huge network can be visualized at different resolutions, offering an intuitive
solution for network analysis and navigation.

     The fast growing social media has spawn novel lines of research on community detection.
The first line focuses on scaling up community detection methods to handle networks of colossal
sizes (Flake et al., 2000; Gibson et al., 2005; Dourisboure et al., 2007; Andersen and Lang, 2006).
This is because many well-studied approaches in social sciences were not designed to handle sizes
of social media network data.

     The second line of research deals with networks of heterogeneous entities and interac-
tions (Zeng et al., 2002; Java et al., 2008; Tang et al., 2008, 2009). Take YouTube as an example.
A network in YouTube can encompass entities like users, videos, and tags; entities of different types
can interact with each other. In addition, users can communicate with each other in distinctive ways,
e.g., connecting to a friend, leaving a comment, sending a message, etc. These distinctive types of
entities or interactions form heterogeneous networks in social media. With a heterogeneous network, we can investigate how communities in one type of entities correlate with those in another
type and how to figure out the hidden communities among heterogeneous interactions.

     The third line of research considers the temporal development of social media net-
works. In essence, these networks are dynamic and evolve with changing community memberships (Backstrom et al., 2006; Palla et al., 2007; Asur et al., 2007; Tang et al., forthcoming). For instance, the number of active users in Facebook has grown from 14 million in 2005[5] to 500 million
as in 2010[6]. As a network evolves, we can study how discovered communities are kept abreast with
its growth and evolution, what temporal interaction patterns are there, and how these patterns can
help identify communities.

[5]http://www.nytimes.com/2005/05/26/business/26sbiz.html
[6]http://www.facebook.com/press/info.php?statistics

Figure 1.4: Link Prediction. Edges $e(2, 4)$ and $e(8, 9)$ are likely links.

1.4.4   CLASSIFICATION AND RECOMMENDATION

Classification and recommendation tasks are common in social media applications. A successful social media site often requires a sufficiently large population. As personalized recommendations can help enhance user experience, it is critical for the site to provide customized recommendations (e.g., friends, communities, tags) to encourage more user interactions with better experience. Classification can help recommendation. For example, one common feature on many social networking sites is *friend recommendation* that suggests a list of friends that a user might know. Since it is hard for a user to figure out who is available on a site, an effective way of expanding one's friendship network is to automatically recommend connections. This problem is known as *link prediction* (Liben-Nowell and Kleinberg, 2007). Essentially, it predicts which nodes are likely to get connected given a social network. The output can be a (ranked) list of links.

Figure 1.4 is an example of link prediction. The given network is on the left side. Based on the network structure, link prediction generates a list of connections that are most likely. In our example, two connections are suggested: $e(2, 4)$ and $e(8, 9)$, resulting in a network on the right in which dashed lines are the predicted links. If a network involves more than one type of entity (say, user-group subscription network, user-item shopping relationship, user-movie ratings, etc.), the recommendation becomes a *collaborative filtering* problem (Breese et al., 1998).

There are other tasks that also involve the utilization of social networks. For instance, given a social network and some user information (interests, preferences, or behaviors), we can infer the information of other users within the same network. Figure 1.5 presents an example of *network-based classification*. The classification task here is to know whether an actor is a smoker or a non-smoker (indicated by + and −, respectively). Users with unknown smoking behavior are colored in yellow with question marks. By studying connections between users, it is possible to infer the behavior of those unknown users as shown on the right. In general, social media offers rich user information in terms of interactions. Consequently, solutions that allow to leverage this kind of network information are desirable in our effort to harness the predictive power of social media.

1.4.5   PRIVACY, SPAM AND SECURITY

Social media entails the socialization between humans. Naturally, *privacy* is an inevitable and sensitive topic. Many social media sites (e.g., Facebook, Google Buzz) often find themselves as the subjects in heated debates about user privacy. It is also not a task that can be dealt with lightly. For example,

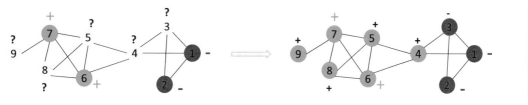

Figure 1.5: Network-based Classification

as shown in (Backstrom et al., 2007), in some cases, social network data, without user identity but simply describing their interactions, can allow an adversary to learn whether there exist connections between specific targeted pairs of users. More often than not, simple anonymization of network nodes cannot protect privacy as one might hope for. For example, this privacy concern was attributed to the cancellation of Nexflix Prize Sequel[7] in 2010.

Spam is another issue that causes grave concerns in social media. In blogosphere, spam blogs (a.k.a., splogs) (Kolari et al., 2006a,b) and spam comments have cropped up. These spams typically contain links to other sites that are often disputable or otherwise irrelevant to the indicated content or context. Identifying these spam blogs becomes increasingly important in building a blog search engine. Attacks are also calling for attentions in social tagging systems (Ramezani et al., 2008). Social media often involves a large volume of private information. Some spammers use fake identifiers to obtain user private information on social networking sites. Intensified research is in demand on building a secure social computing platform as it is critical in turning social media sites into a successful marketplace[8].

1.5   SUMMARY

Social media mining is a young and vibrant field with many promises. Opportunities abound. For example, the link prediction technique in (Liben-Nowell and Kleinberg, 2007) achieves significantly better performance than random prediction, but its absolute accuracy is far from satisfactory. For some other tasks, we still need to relax and remove unrealistic assumptions in state-of-the-art algorithms.

Social media has kept surprising us with its novel forms and variety. Examples include the popular microblogging service Twitter that restricts each tweet to be of at most 140 characters, and the location-based updating service Foursquare. An interesting trend is that social media is increasingly blended into the physical world with recent mobile technologies and smart phones. In other words, as social media is more and more weaved into human beings' daily lives, the divide between the virtual and physical worlds blurs, a harbinger of the convergence of mining social media and mining the reality (Mitchell, 2009).

The remainder of this lecture consists of four chapters. Chapter 2 is a foundation of many social computing tasks, introducing nodes, ties, centrality analysis, and influence modeling. Chapter 3

---

[7]http://blog.netflix.com/2010/03/this-is-neil-hunt-chief-product-officer.html
[8]On August 13, 2010, Delta Airlines launched their flight booking service on Facebook.

illustrates representative approaches to community detection and discusses issues of evaluation. Chapter 4 expands further on community detection in social media networks in presence of many types of heterogeneity. And Chapter 5 contains two social media mining tasks (community evolution and behavior classification) that demonstrate how community detection can help accomplish other social computing tasks. We hope that this lecture can help readers to appreciate a good body of existing techniques and gain better insights in their efforts to build socially-intelligent systems and harness the power of social media.

CHAPTER 2

Nodes, Ties, and Influence

In a social network, *nodes* are usually not independent of each other; they are connected by *ties* (or *edges*, *links*). When nodes are connected, they could influence each other. In a broader sense, influence is a form of contagion that moves in a network of connected nodes. It can be amplified or attenuated. In this chapter, we discuss importance of nodes, strengths of ties, and influence modeling.

2.1 IMPORTANCE OF NODES

It is natural to question which nodes are important among a large number of connected nodes. Centrality analysis provides answers with measures that define the importance of nodes. We introduce classical and commonly used ones (Wasserman and Faust, 1994): *degree centrality*, *closeness centrality*, *betweenness centrality*, and *eigenvector centrality*. These centrality measures capture the importance of nodes in different perspectives.

Degree Centrality The importance of a node is determined by the number of nodes adjacent to it. The larger the degree of one node, the more important the node is. Node degrees in most social media networks follow a power law distribution, i.e., a very small number of nodes have an extremely large number of connections. Those high-degree nodes naturally have more impact to reach a larger population than the remaining nodes within the same network. Thus, they are considered to be more important.

The degree centrality is defined[1] as

$$C_D(v_i) = d_i = \sum_j A_{ij}. \tag{2.1}$$

When one needs to compare two nodes in different networks, a normalized degree centrality should be used,

$$C'_D(v_i) = d_i/(n-1). \tag{2.2}$$

Here, n is the number of nodes in a network. It is the proportion of nodes that are adjacent to node v_i. Take node 1 in the toy network in Figure 1.1 as an example. Its degree centrality is 3, and its normalized degree centrality is $3/(9-1) = 3/8$.

Closeness Centrality A central node should reach the remaining nodes more quickly than a non-central node. Closeness centrality measures how close a node is to all the other nodes. The

---

[1]Refer to Table 1.4 for symbol definitions.

measure involves the computation of the average distance of one node to all the other nodes:

$$D_{avg}(v_i) = \frac{1}{n-1} \sum_{j \neq i}^{n} g(v_i, v_j), \tag{2.3}$$

where n is the number of nodes, and $g(v_i, v_j)$ denotes the geodesic distance between nodes v_i and v_j. The average distance can be regarded as a measure of how long it will take for information starting from node v_i to reach the whole network. Conventionally, a node with higher centrality is more important. Thus, the closeness centrality is defined as a node's inverse average distance,

$$C_C(v_i) = \left[\frac{1}{n-1} \sum_{j \neq i}^{n} g(v_i, v_j) \right]^{-1} = \frac{n-1}{\sum_{j \neq i}^{n} g(v_i, v_j)} \tag{2.4}$$

The pairwise distance between nodes in Figure 1.1 is exemplified in Table 2.1.

Table 2.1: Pairwise geodesic distance

Node	1	2	3	4	5	6	7	8	9
1	0	1	1	1	2	2	3	3	4
2	1	0	1	2	3	3	4	4	5
3	1	1	0	1	2	2	3	3	4
4	1	2	1	0	1	1	2	2	3
5	2	3	2	1	0	1	1	1	2
6	2	3	2	1	1	0	1	1	2
7	3	4	3	2	1	1	0	1	1
8	3	4	3	2	1	1	1	0	2
9	4	5	4	3	2	2	1	2	0

The closeness centrality of nodes 3 and 4 are

$$C_C(3) = \frac{9-1}{1+1+1+2+2+3+3+4} = 8/17 = 0.47,$$
$$C_C(4) = \frac{9-1}{1+2+1+1+1+2+2+3} = 8/13 = 0.62.$$

As $C_C(4) > C_C(3)$, we conclude that node 4 is more central than node 3. This is consistent with what happens in Figure 1.1.

Betweenness Centrality Node betweenness counts the number of shortest paths in a network that will pass a node. Those nodes with high betweenness play a key role in the communication

Table 2.2: $\sigma_{st}(4)/\sigma_{st}$

	s = 1	s = 2	s = 3
t = 5	1/1	2/2	1/1
t = 6	1/1	2/2	1/1
t = 7	2/2	4/4	2/2
t = 8	2/2	4/4	2/2
t = 9	2/2	4/4	2/2

Table 2.3: $\sigma_{st}(5)/\sigma_{st}$

	s = 1	s = 2	s = 3	s = 4
t = 7	1/2	2/4	1/2	1/2
t = 8	1/2	2/4	1/2	1/2
t = 9	1/2	2/4	1/2	1/2

within the network. The betweenness centrality of a node is defined as

$$C_B(v_i) = \sum_{v_s \neq v_i \neq v_t \in V, s < t} \frac{\sigma_{st}(v_i)}{\sigma_{st}}, \qquad (2.5)$$

where σ_{st} is the total number of shortest paths between nodes v_s and v_t, and $\sigma_{st}(v_i)$ is the number of shortest paths between nodes v_s and v_t that pass along node v_i.

For instance, $\sigma_{19} = 2$ in the network in Figure 1.1, as there are two shortest paths between nodes 1 and 9: 1-4-5-7-9 and 1-4-6-7-9. Consequently, $\sigma_{19}(4) = 2$, and $\sigma_{19}(5) = 1$. It can be shown that $C_B(4) = 15$. As all shortest paths from $\{1, 2, 3\}$ to $\{5, 6, 7, 8, 9\}$ have to pass node 4. The number of shortest paths and those passing node 4 are listed in Table 2.2, where s and t are the source node and the target node respectively. Hence, we have $C_B(4) = 15$. As for node 5, $C_B(5) = 6$. The related statistics to calculate the betweenness are given in Table 2.3. All the shortest paths from node $\{1, 2, 3, 4\}$ to nodes $\{7, 8, 9\}$ have to pass either node 5 and 6. As there are in total $3 \times 4 = 12$ pairs, and each pair has a 50% chance to pass node 5. Hence, $C_B(5) = 12 \times 0.5 = 6$. The betweenness of other nodes are shown in Table 2.4.

Table 2.4: Betweenness Centrality

Node	1	2	3	4	5	6	7	8	9
Betweenness	3	0	3	15	6	6	7	0	0

Basically, for each pair of nodes v_s and v_t, we compute the probability that their shortest paths will pass node v_i. Since the maximum value of $C_B(i)$ in an undirected network can be

$$\binom{n-1}{2} = (n-1)(n-2)/2,$$

we can normalize the betweenness centrality as

$$C'_B(i) = \frac{C_B(i)}{(n-1)(n-2)/2}. \tag{2.6}$$

However, the computation of shortest paths between all pairs is computationally prohibitive for large-scale networks. It takes at least $O(nm)$ time to compute (Brandes, 2001), where n is the number of nodes, and m the number of edges. An algorithm to compute betweenness centrality is included in appendix B.

Eigenvector Centrality The underlying idea beneath eigenvector centrality is that one's importance is defined by his friends' importance. In other words, if one has many important friends, he should also be important. In particular,

$$C_E(v_i) \propto \sum_{v_j \in N_i} A_{ij} C_E(v_j).$$

Let $\mathbf{x}$ denote the eigenvector centrality of node from v_1 to v_n. The above equation can be written as in a matrix form:

$$\mathbf{x} \propto A\mathbf{x}.$$

Equivalently, we can write $\mathbf{x} = \frac{1}{\lambda} A\mathbf{x}$, where λ is a constant. It follows that

$$A\mathbf{x} = \lambda \mathbf{x}.$$

Thus $\mathbf{x}$ is an eigenvector of the adjacency matrix A. Eigenvector centrality is defined as the principal eigenvector of the adjacency matrix defining the network.

Indeed, Google's Pagerank (Page et al., 1999) is a variant of the eigenvector centrality. In Pagerank, a transition matrix is constructed based on the adjacency matrix by normalizing each column to a sum of 1:

$$\widetilde{A}_{ij} = A_{ij} / \sum_i A_{ij}.$$

In the transition matrix $\widetilde{A}$, an entry $\widetilde{A}_{ij}$ denotes the probability of transition from node v_j to node v_i. In the context of web surfing, it denotes the probability of one user clicking on a web page (node v_i) after browsing current web page (node v_j) by following the link $e(v_j, v_i)$[2]. For example, given the adjacency matrix in Table 1.3, we have a transition matrix as shown in Table 2.5.

Pagerank scores correspond to the top eigenvector of the transition matrix $\widetilde{A}$. It can be computed by the power method, i.e., repeatedly left-multiplying a non-negative vector $\mathbf{x}$ with $\widetilde{A}$.

---

[2]A damping factor might be added to the transition matrix to account for the probability that a user jumps from one page to another web page rather than following links.

Table 2.5: Column-Normalized Adjacency Matrix								
Node	1	2	3	4	5	6	7	8	9
1	0	1/2	1/3	1/4	0	0	0	0	0
2	1/3	0	1/3	0	0	0	0	0	0
3	1/3	1/2	0	1/4	0	0	0	0	0
4	1/3	0	1/3	0	1/4	1/4	0	0	0
5	0	0	0	1/4	0	1/4	1/4	1/3	0
6	0	0	0	1/4	1/4	0	1/4	1/3	0
7	0	0	0	0	1/4	1/4	0	1/3	1
8	0	0	0	0	1/4	1/4	1/4	0	0
9	0	0	0	0	0	0	1/4	0	0

Suppose we start from $\mathbf{x}^{(0)} = \mathbf{1}$, then $x^{(1)} \propto \tilde{A}x^{(0)}$, $x^{(2)} \propto \tilde{A}x^{(1)}$, etc. Typically, the vector $\mathbf{x}$ is normalized to the unit length. Below, we show the values of $\mathbf{x}$ in the first seven iterations.

$$\mathbf{x}^{(0)} = \begin{bmatrix} 1 \\ 1 \\ 1 \\ 1 \\ 1 \\ 1 \\ 1 \\ 1 \\ 1 \end{bmatrix}, \mathbf{x}^{(1)} = \begin{bmatrix} 0.33 \\ 0.21 \\ 0.33 \\ 0.36 \\ 0.33 \\ 0.33 \\ 0.57 \\ 0.23 \\ 0.08 \end{bmatrix}, \mathbf{x}^{(2)} = \begin{bmatrix} 0.32 \\ 0.23 \\ 0.32 \\ 0.41 \\ 0.41 \\ 0.41 \\ 0.34 \\ 0.32 \\ 0.15 \end{bmatrix}, \mathbf{x}^{(3)} = \begin{bmatrix} 0.32 \\ 0.21 \\ 0.32 \\ 0.41 \\ 0.39 \\ 0.39 \\ 0.45 \\ 0.28 \\ 0.08 \end{bmatrix},$$

$$\mathbf{x}^{(4)} = \begin{bmatrix} 0.32 \\ 0.21 \\ 0.32 \\ 0.41 \\ 0.41 \\ 0.41 \\ 0.37 \\ 0.31 \\ 0.11 \end{bmatrix}, \mathbf{x}^{(5)} = \begin{bmatrix} 0.31 \\ 0.21 \\ 0.31 \\ 0.41 \\ 0.40 \\ 0.40 \\ 0.42 \\ 0.30 \\ 0.10 \end{bmatrix}, \mathbf{x}^{(6)} = \begin{bmatrix} 0.31 \\ 0.21 \\ 0.31 \\ 0.41 \\ 0.41 \\ 0.41 \\ 0.39 \\ 0.31 \\ 0.11 \end{bmatrix}, \mathbf{x}^{(7)} = \begin{bmatrix} 0.31 \\ 0.21 \\ 0.31 \\ 0.41 \\ 0.40 \\ 0.40 \\ 0.41 \\ 0.30 \\ 0.10 \end{bmatrix}.$$

After convergence, we have the Pagerank scores for each node listed in Table 2.6. Based on eigenvector centrality, nodes {4, 5, 6, 7} share similar importance.

With large-scale networks, the computation of centrality measures can be expensive except for degree centrality and eigenvector centrality. Closeness centrality, for instance, involves the computation of all the pairwise shortest paths, with time complexity of $O(n^2)$ and space complexity of

Table 2.6: Eigenvector Centrality								
1	2	3	4	5	6	7	8	9
0.31	0.20	0.31	0.41	0.41	0.41	0.41	0.31	0.10

$O(n^3)$ with the Floyd-Warshall algorithm (Floyd, 1962) or $O(n^2 \log n + nm)$ time complexity with Johnson's algorithm (Johnson, 1977). The betweenness centrality requires $O(nm)$ computational time following (Brandes, 2001). The eigenvector centrality can be computed with less time and space. Using a simple power method (Golub and Van Loan, 1996) as we did above, the eigenvector centrality can be computed in $O(m\ell)$ where ℓ is the number of iterations. For large-scale networks, efficient computation of centrality measures is critical and requires further research.

2.2   STRENGTHS OF TIES

Many studies treat a network as an unweighted graph or a boolean matrix as we introduced in Chapter 1. However, in practice, those ties are typically not of the same strength. As defined by Granovetter (1973), "the strength of a tie is a (probably linear) combination of the amount of time, the emotional intensity, the intimacy (mutual confiding), and the reciprocal services which characterize the tie." Interpersonal social networks are composed of strong ties (close friends) and weak ties (acquaintances). Strong ties and weak ties play different roles for community formation and information diffusion. It is observed by Granvoetter that occasional encounters with distant acquaintances can provide important information about new opportunities for job search.

Owing to the lowering of communication barrier in social media, it is much easier for people to connect online and interact. Consequently, some users might have thousands of connections, which is rarely true in the physical world. Among one's many online friends, only some are "close friends," while others are kept just as contacts in an address book. Thus, it is imperative to estimate tie strengths (Gilbert and Karahalios, 2009), given a social media network. There exist three chief approaches to this task: 1) analyzing network topology, 2) learning from user attributes, and 3) learning from user activities.

2.2.1   LEARNING FROM NETWORK TOPOLOGY

Intuitively, the edges connecting two different communities are called "bridges". An edge is a *bridge* if its removal results in the disconnection of the two terminal nodes. Bridges in a network are weak ties. For instance, in the network in Figure 2.1, edge $e(2, 5)$ is a weak tie. If the edge is removed, nodes 2 and 5 are not connected anymore. However, in real-world networks, such bridges are not common. It is more likely that after the removal of one edge, the terminal nodes are still connected through alternative paths as shown in Figure 2.2. If we remove edge $e(2, 5)$ from the network, nodes 2 and 5 are still connected through nodes 8, 9, and 10. The strength of a tie can be calibrated by the length of an alternative shortest path between the end points of the edge. If there is no path between the two terminal nodes, the geodesic distance is $+\infty$. The larger the geodesic distance between terminal

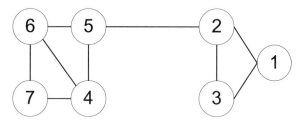

Figure 2.1: A network with $e(2, 5)$ being a bridge. After its removal, nodes 2 and 5 become disconnected.

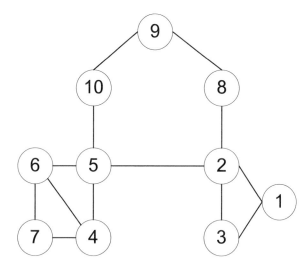

Figure 2.2: A network in which the removal of $e(2, 5)$ increases the geodesic distance between nodes 2 and 5 to 4.

nodes after the edge removal, the weaker the tie is. For instance, in the network in Figure 2.2, after removal of edge $e(2, 5)$, the geodesic distance $d(2, 5) = 4$. Comparatively, $d(5, 6) = 2$ if the edge $e(5, 6)$ is removed. Consequently, edge $e(5, 6)$ is a stronger tie than $e(2, 5)$.

Another way to measure tie strengths is based on the neighborhood overlap of terminal nodes (Onnela et al., 2007; Easley and Kleinberg, 2010). Let N_i denote the friends of node v_i. Given a link $e(v_i, v_j)$, the neighborhood overlap is defined as

$$overlap(v_i, v_j) = \frac{\text{number of shared friends of both } v_i \text{ and } v_j}{\text{number of friends who are adjacent to at least } v_i \text{ or } v_j}$$
$$= \frac{|N_i \cap N_j|}{|N_i \cup N_j| - 2}.$$

We have -2 in the denominator just to exclude v_i and v_j from the set $N_i \cup N_j$. Typically, the larger the overlap, the stronger the connection. For example, it was reported in (Onnela et al., 2007) that the neighborhood overlap is positively correlated with the total number of minutes spent by two persons in a telecommunication network. As for the example in Figure 2.2, $N_2 = \{1, 3, 5, 8\}$, $N_5 = \{2, 4, 6, 10\}$. Since $N_2 \cap N_5 = \phi$, we have $overlap(2, 5) = 0$, indicating a weak tie between them. On the contrary, the neighborhood overlap between nodes 5 and 6 is

$$overlap(5, 6) = \frac{|\{4\}|}{|\{2, 4, 5, 6, 7, 10\}| - 2} = 1/4.$$

Thus, edge $e(5, 6)$ is a stronger tie than $e(2, 5)$.

2.2.2   LEARNING FROM USER ATTRIBUTES AND INTERACTIONS

In reality, people interact frequently with very few of those "listed" online friends, empirically verified by several studies. One case study was conducted by Huberman et al. (2009) on the Twitter site. Twitter is a microblogging platform that users can tweet, i.e., post a short message to update their status or share some links. They can also post directly to another user. One user in Twitter can become a follower of another user without the followee's confirmation. Consequently, the follower network in Twitter is directional. Huberman et al. define a Twitter user's friend as a person whom the user has directed at least two posts to. It is observed that Twitter users have a very small number of friends compared to the number of followers and followees they declare. The friendship network is more influential in driving Twitter usage rather than the denser follower-followee network. People with many followers or followees do not necessarily post more than those with fewer followers or followees.

Gilbert and Karahalios (2009) predict the strength of ties based on social media data. In particular, they use Facebook as a testbed and collect various attribute information of user interactions. Seven types of information are collected: predictive intensity variables (say, friend-initiated posts, friends' photo comments), intimacy variables (say, participants' number of friends, friends' number of friends), duration variable (days since first communication), reciprocal service variables (links exchanged by wall post, applications in common), structural variables (number of mutual friends), emotional support variables (positive/negative emotion words in one user's wall or inbox), and social distance variables (age difference, education difference). The authors build a linear predictive model from these variables for classifying the tie strengths based on the data collected from the user survey response, and they show that the model can distinguish between strong and weak ties with over 85% accuracy.

Xiang et al. (2010) propose to represent the strengths of ties using numerical weights instead of just "strong" and "weak" ties. They treat the tie strengths as latent variables. It is assumed that the similarity in user profiles determines the strength of their relationship, which in turn determines user interaction. The user profile features include whether two users attend the same school, work at the same company, live in the same location, etc. User interactions, depending on the social media site and available information, can include features such as whether two have established a connection, whether one writes a recommendation for the other, and so on. The authors show that the strengths can be learned by optimizing the joint probability given user profiles and interaction information. It is demonstrated that the learned strengths can help improve classification and recommendation.

2.2.3   LEARNING FROM SEQUENCE OF USER ACTIVITIES

The previous two approaches mainly focus on static networks or network snapshots. Given a stream of user activities, it is possible to learn the strengths of ties based on the influence probability. As this involves a sequence of user activities, the temporal information has to be considered.

Kossinets et al. (2008) study how information is diffused in communication networks. They mark the latest information available to each actor at each timestamp. It is observed that a lot of information diffusion violates the "triangle inequality". There is one shortest path between two nodes based on network topology, but information does not necessarily propagate following the shortest path. Alternatively, the information diffuses following certain paths that reflect the roles of actors and the true communication pattern. *Network backbones* are defined to be those ties that are likely to bear the task of propagating the timely information.

At the same time, one can learn the strengths of ties by studying how users influence each other. Logs of user activities (e.g., subscribing to different interest groups, commenting on a photo) (Goyal et al., 2010; Saito et al., 2010) may be available. By learning the probabilities that one user influences his friends over time, we can have a clear picture of which ties are more important. This kind of approach hinges on the adopted influence model, which will be introduced in the next section.

2.3   INFLUENCE MODELING

Influence modeling is one of the fundamental questions in order to understand the information diffusion, spread of new ideas, and word-of-mouth (viral) marketing. Among various kinds of influence models, two basic ones that garner much attention are the *linear threshold model* and the *independent cascade model* (Kempe et al., 2003). For simplicity, we assume that one actor is active if he adopts a targeted action or chooses his preference. The two models share some properties in the diffusion process:

- A social network is represented a *directed* graph, with each actor being one node;

- Each node is started as active or inactive;

- A node, once activated, will activate his neighboring nodes;

- Once a node is activated, this node cannot be deactivated[3].

---

[3]This assumption may be unrealistic in some diffusion process. But both models discussed next can be generalized to handle more realistic cases (Kempe et al., 2003).

2.3.1   LINEAR THRESHOLD MODEL (LTM)

The threshold model dates back to 1970s (Granovetter, 1978). It states that an actor would take an action if the number of his friends who have taken the action exceeds a certain threshold. In his seminal work, Thomas C. Schelling essentially used a variant of the threshold model to show that a small preference for one's neighbors to be of the same color could lead to *racial segregation* (Schelling, 1971). Many variants of the threshold model have been studied. Here, we introduce one linear threshold model.

In a linear threshold model, each node v chooses a threshold θ_v randomly from a uniform distribution in an interval between 0 and 1. The threshold θ_v represents the fraction of friends of v to be active in order to activate v. Suppose that a neighbor w can influence node v with strength $b_{w,v}$. Without loss of generality, we assume that

$$\sum_{w \in N_v} b_{w,v} \leq 1.$$

Given randomly assigned thresholds to all nodes, and an initial active set A_0, the diffusion process unfolds deterministically. In each discrete step, all nodes that were active in the previous step remain active. The nodes satisfying the following condition will be activated as

$$\sum_{w \in N_v, w \text{ is active}} b_{w,v} \geq \theta_v. \tag{2.7}$$

The process continues until no further activation is possible.

Take the network in Figure 1.1 as an example. We assume the network is directed. If a network is directed, the weights $b_{w,v}$ and $b_{v,w}$ between nodes v and w are typically different in an influence

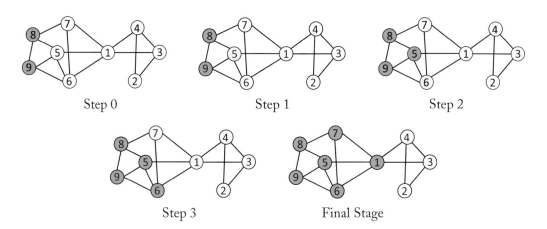

Figure 2.3: A diffusion process following the *linear threshold model*. Green nodes are the active ones, and yellow nodes the newly activated ones.

model. For simplicity, we assume $b_{w,v} = 1/k_v$ and $b_{v,w} = 1/k_w$, and the threshold for each node is 0.5. Suppose we start from two activated nodes 8 and 9. Figure 2.3 shows the diffusion process for the linear threshold model. In the first step, actor 5, with two of its neighbors being active, receives weights $b_{8,5} + b_{9,5} = 1/3 + 1/3 = 2/3$ from activated nodes, larger than its threshold 0.5. Thus, actor 5 is activated. In the second step, node 6, with two of its friends being active, will be active. In a similar vein, nodes 7 and 1 will be active in the third step. After that, no more nodes can be activated. Hence, the diffusion process terminates after nodes 1, 5, 6, 7, 8, 9 become active.

Note that the linear threshold model presented in (Kempe et al., 2003) assumes the node thresholds are randomly sampled from a uniform distribution of an interval [0, 1] before the diffusion process starts. Once the thresholds are fixed, the diffusion process is determined. Many studies also hard-wire all thresholds at a fixed value like 0.5, but this kind of threshold model does not satisfy *submodularity* that is discussed in Section 2.3.3.

2.3.2   INDEPENDENT CASCADE MODEL (ICM)

The *independent cascade model* borrows the idea from interacting particle systems and probability theory. Different from the linear threshold model, the independent cascade model focuses on the sender's rather than the receiver's view. In the independent cascade model, a node w, once activated at step t, has *one* chance to activate *each of its neighbors*. For a neighboring node (say, v), the activation succeeds with probability $p_{w,v}$. If the activation succeeds, then v will become active at step $t + 1$. In the subsequent rounds, w will not attempt to activate v anymore. The diffusion process, the same as that of the linear threshold model, starts with an initial activated set of nodes, then continues until no further activation is possible.

Let us now apply ICM to the network in Figure 2.3. We assume $p_{w,v} = 0.5$ for all edges in the network, i.e., a node, once activated, will activate his inactive neighbors with a 50% chance. If nodes 8 and 9 are activated, Figure 2.4 shows the diffusion process following ICM. Starting from the initial stage with nodes 8 and 9 being active, ICM choose their neighbors and activate them with success rate equaling to $p_{v,w}$. At Step 1, nodes 8 and 9 will try to activate nodes 5, 6, and 7. Suppose the activation succeeds for nodes 5 and 7, but fails for node 6. Now, given two newly activated nodes 5 and 7, we activate their neighbors by flipping a coin. Say, both nodes 1 and 6 become active in Step 2. Because all the neighboring nodes of actor 6 is already active, we only consider those inactive friends of node 1 in Step 3. That is, following ICM, we will flip a coin to activate nodes 3 and 4 respectively. Assuming only actor 4 becomes active successfully, then in the next step, we consider the neighboring nodes of the newly activated node 4. Unfortunately, neither of them succeeds. Thus, the diffusion process stops with nodes 1, 4, 5, 6, 7, 8, 9 being active. Note that ICM activates one node with certain success rate. Thus, we might get a different result for another run.

Clearly, both the linear threshold model and the independent cascade model capture the information diffusion in a certain aspect (Gruhl et al., 2004), but demonstrate a significant difference. LTM is receiver-centered. By looking at all the neighboring nodes of one node, it determines whether to activate the node based on its threshold. ICM, on the contrary, is sender-centered. Once a node

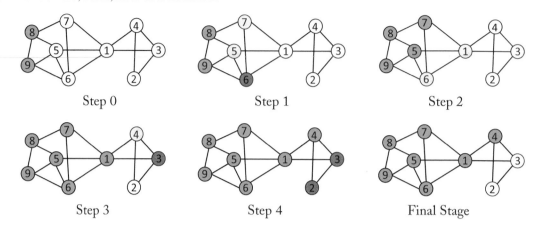

Figure 2.4: A diffusion process following the *independent cascade model*. Green nodes are the active ones, yellow nodes the newly activated ones, and red nodes those ones that do not succeed in activation.

is activated, it tries to activate all its neighboring nodes. LTM's activation depends on the whole neighborhood of one node, where ICM, as indicated by its name, activates each of its neighbors independently. Another difference is that LTM, once the thresholds are sampled, the diffusion process is determined. ICM, however, varies depending on the cascading process. Both models involve randomization: LTM randomly selects a threshold for each node at the outset, whereas ICM succeeds activates a neighboring node with probability $p_{w,v}$.

Of course, there are other general influence models. LTM and ICM are two basic models that are used to study influence and information diffusion. Because both are *submodular*, it warrants a fast approximation of the influence maximization with a theoretical guarantee (Kempe et al., 2003).

2.3.3   INFLUENCE MAXIMIZATION

The influence maximization problem can be formulated as follows:

> Given a network and a parameter k, which k nodes should be selected to be in the activation set B in order to maximize the influence in terms of active nodes at the end? Let $\sigma(B)$ denote the expected number of nodes that can be influenced by B, the optimization problem can be formulated as

$$\max_{B \subseteq V} \sigma(B) \; s.t. \; |B| \leq k. \tag{2.8}$$

This problem is closely related to the *viral marketing* problem (Domingos and Richardson, 2001; Richardson and Domingos, 2002). Other variants include cost-effective blocking in problems such as water contamination and blog cascade detection (Leskovec et al., 2007b), and personalized blog selection and monitoring (El-Arini et al., 2009). Here, we consider the simplest form of the viral marketing problem, i.e., excluding varying cost or types of strategies for marketing different nodes. Maximizing the influence, however, is a NP-hard problem under either type of the diffusion model (LTM or ICM).

One natural approach is greedy selection. It works as follows. Starting with $B = \phi$, we evaluate $\sigma(v)$ for each node, and pick the node with maximum σ as the first node v_1 to form $B = \{v_1\}$. Then, we select a node which will increase $\sigma(B)$ most if the node is included in B. Essentially, we greedily find a node $v \in V \setminus B$ such that

$$v = \arg \max_{v \in V \setminus B} \sigma(B \cup \{v\}), \tag{2.9}$$

or equivalently,

$$v = \arg \max_{v \in V \setminus B} \sigma(B \cup \{v\}) - \sigma(B). \tag{2.10}$$

Suppose our budget allows to convert at most two nodes in the network in Figure 2.3. A greedy approach will first pick the node with the maximum expected influence population (i.e., node 1), and then pick the second node that will increase the expected influence population maximally given node 1 is active. Such a greedy approach has been observed to yield better performance than selecting the top k nodes with the maximum node centrality (Kempe et al., 2003). Moreover, it is proved that the greedy approach gives a solution that is at least 63% of the optimal. That is because the influence function $\sigma(\cdot)$ is *monotone* and *submodular* under both the linear threshold model and the independent cascade model.

- A set function $\sigma(\cdot)$ mapping from a set to a real value is *monotone* if

$$\sigma(S \cup \{v\}) \geq \sigma(S).$$

- A set function $\sigma(\cdot)$ is *submodular* if it satisfying the *diminishing returns* property: the marginal gain from adding an element to a set S is no less than the marginal gain from adding the same element to a superset of S. Formally, given two sets S and T such that $S \subseteq T$, $\sigma(\cdot)$ is submodular if it satisfies the property below:

$$\sigma(S \cup \{v\}) - \sigma(S) \geq \sigma(T \cup \{v\}) - \sigma(T). \tag{2.11}$$

Suppose we have a set function as follows:

$$\tilde{\sigma}(B) = 1 + \frac{1}{2} + \frac{1}{3} + \cdots + \frac{1}{|B|}.$$

$\tilde{\sigma}(\cdot)$ is monotone because

$$\tilde{\sigma}(B \cup \{v\}) = \begin{cases} 1 + \frac{1}{2} + \frac{1}{3} + \cdots + \frac{1}{|B|} + \frac{1}{|B|+1} = \tilde{\sigma}(B) + \frac{1}{|B|+1} & \text{if } v \notin B \\ \tilde{\sigma}(B) & \text{if } v \in B \end{cases}$$

$$\geq \tilde{\sigma}(B)$$

$\tilde{\sigma}$ is also submodular. Given $S \subseteq T$, it follows that $|S| \leq |T|$. Thus, $\sigma(S \cup \{v\}) - \sigma(S) = \frac{1}{|S|+1} \geq \frac{1}{|T|+1} = \sigma(T \cup \{v\}) - \sigma(T)$ if $v \notin S, v \notin T$. Other cases can also be verified.

Theorem 2.1   *(Nemhauser et al., 1978) If $\sigma(\cdot)$ is a monotone, submodular set function and $\sigma(\phi) = 0$, then the greedy algorithm finds a set B_G, such that*

$$\sigma(B_G) \geq (1 - 1/e) \cdot max_{|B|=k}\sigma(B).$$

Here $1 - 1/e \approx 63\%$. As the expected number of influenced nodes under LTM or ICM is a submodular function of nodes, the greedy algorithm can output a reasonable (at least 63% optimal) solution to the NP-hard influence maximization problem. Given selected nodes B, we need to find a new node that maximally increases σ. It can still be time-consuming to evaluate $\sigma(B \cup v)$ for all possible choices of v. Kempe et al. (2003), for instance, simulates the diffusion process many times in order to estimate $\sigma(B \cup v)$, as we did in Figures 2.3 and 2.4. When a network has millions of nodes, the evaluation of all possible choices of $\sigma(B \cup v)$ becomes a challenge for efficient computation.

Leskovec et al. (2007b) suggest that the number of evaluations can be reduced dramatically by exploiting the submodularity. Following the greedy approach, in each iteration, we aim to find a node with the maximal marginal gain (i.e., $\sigma(B \cup \{v\}) - \sigma(B)$) and add it to B. Following Eq. (2.11), the marginal gain of adding a node v to a selected set B can only decrease after we expand B. Suppose we evaluate the marginal gain of a node v in one iteration and find out the gain is Δ. Then, those nodes whose marginal gain is less than Δ in the *previous* iteration should not be considered for evaluation because their marginal gains can only decrease. We will not add them to B since v is at least a better choice. Thus, many unnecessary evaluations of $\sigma(\cdot)$ can be avoided. In practice, we can simply maintain a priority queue to first evaluate those nodes whose marginal gains are larger in previous iterations. This has been reported to improve a straightforward implementation of the greedy approach by up to 700 times. More efficient heuristics are explored in (Chen et al., 2009, 2010).

In the meantime, some researchers question about the importance of influential users. Watts and Dodds (2007) reported in their study that "large-scale changes in public opinion are not driven by highly influential people who influence everyone else but rather by easily influenced people influencing other easily influenced people". According to them, when a population is ready to embrace a trend change, anyone can start the cascade. The so-called influentials may be just "accidental influential" because they are the early adopters. It remains an open question to model the formation of public opinions and trend change.

2.3.4   DISTINGUISHING INFLUENCE AND CORRELATION

It has been widely observed that user attributes and behaviors tend to correlate with their social networks (Singla and Richardson, 2008). We can perform a simple test to check whether there is any correlation associated with a social network (Easley and Kleinberg, 2010). Suppose we have a binary attribute with each node (say, whether or not a smoker). If the attribute is correlated with a social network, we expect actors sharing the same attribute value to be positively correlated with social connections. That is, smokers are more likely to interact with other smokers, and non-smokers with non-smokers. Thus, the probability of connections between nodes bearing different attribute values (i.e., a smoker with a non-smoker) should be relatively low. Given a network, we can count the fraction of edges connecting nodes with distinctive attribute values. Then, we compare it to the *expected* probability of such connections if the attribute and the social connections are independent. If the two quantities are significantly different, we conclude the attribute is correlated with the network.

The expected probability of connections between nodes bearing different attribute values can be computed as follows. In a given network, say we observe a p fraction of actors are smokers, and a $(1 - p)$ fraction are non-smokers. If connections are independent of the smoking behavior, one edge is expected to connect two smokers with probability $p \times p$, and two non-smokers with probability $(1 - p) \times (1 - p)$. And the probability of one edge linking a smoker with a non-smoker is $1 - p^2 - (1 - p)^2 = 2p(1 - p)$. Thus, we can perform the following test for correlation[4].

> **Test for Correlation:** If the fraction of edges linking nodes with different attribute values are significantly less than the expected probability, then there is evidence of correlation.

For the network in Figure 2.5, a 4/9 fraction of nodes are smokers and 5/9 are non-smokers. Thus, the expected probability of an edge connecting a smoker and non-smoker is $2 \times 4/9 \times 5/9 = 49\%$. In other words, if the connections are independent of the individual behavior, we would expect almost half of the connections to reside between a smoker and a non-smoker. As seen in the network, the fraction of such connections is only $2/14 = 14\% < 49\%$. Consequently, we conclude this network demonstrates some degree of correlation with respect to the smoking behavior. A more formal way is to conduct a χ^2 test for independence of social connections and attributes (La Fond and Neville, 2010).

It is well known that there exist correlations between behaviors or attributes of adjacent actors in a social network. Three major social processes to explain correlation (Figure 2.6) are: *homophily*, *confounding*, and *influence* (Anagnostopoulos et al., 2008):

- Homophily (McPherson et al., 2001) is a term coined by sociologist in 1950s to explain our tendency to link to others that share certain similarity with us, e.g., age, education level, ethics, interests, etc. In short, "birds of a feather flock together". Homophily assumes the similarity between users breeds connections. This social process is also known as *selection* (Crandall et al., 2008), i.e., people select others who resemble themselves in certain aspects to be friends.

---

[4]Some people also use *homophily* to denote this correlation such as in (Easley and Kleinberg, 2010). Homophily was coined as one of the fundamental social processes to explain correlation observed in a network. In this lecture, we stick to the original meaning of homophily as a social process.

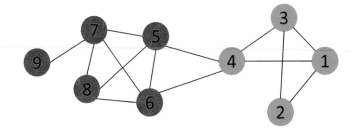

Figure 2.5: A Network Demonstrating Correlations. Red nodes denote non-smokers, and green ones are smokers. If there is no correlation, then the probability of one edge connecting a smoker and a non-smoker is $2 \times 4/9 \times 5/9 = 49\%$.

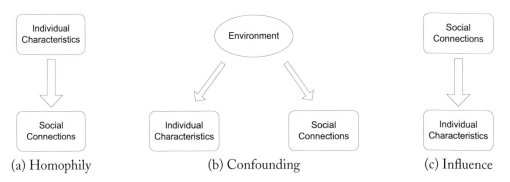

Figure 2.6: Different Social Processes to Explain Correlations observed in Social Networks.

- Correlation between actors can also be forged due to external influences from environment. The environment is referred to as a confounding factor in statistics (Pearl, 2000). In essence, some latent variables might lead to social connections as well as similar individual behaviors. For example, "two individuals living in the same city are more likely to become friends than two random individuals and are also more likely to take pictures of similar scenery and post them on Flickr with the same tag"(Anagnostopoulos et al., 2008).

- A well-known process that causes behavioral correlations between adjacent actors is *influence*, For example, if most of one's friends switch to a mobile company, he might be influenced by his friends and switch to the company as well. In this process, one's social connections and the behavior of his friends affect his decision.

In many studies about influence modeling, influence is determined by timestamps: actor a influences actor b as long as a becomes active at an earlier time than b. But based on a temporal order of user activities, it is hard to conclude whether there is really any influence. The correlation in terms of user behaviors might simply due to the homophily or confounding effect. How can we be sure whether there is any influence in a social network with respect to a behavior? *How to distinguish influence from homophily or confounding?*

Now with the observation of social networks and historical records of user behavior, we might be able to answer. In order to identify whether influence is a factor associated with a social system, one proposed approach is the *shuffle test* (Anagnostopoulos et al., 2008). This test assumes we have access to a social network among users, logs of user behaviors with timestamps. Following the convention in influence modeling (Section 2.3), we say one actor is active if he performs a target action. Based on the user activity log, we can compute a social correlation coefficient. A simple probabilistic model is adopted to measure the social correlation. The probability of one node being active is a logistic function of the number of his active friends as follows:

$$p(a) = \frac{e^{\alpha \ln(a+1)+\beta}}{1 + e^{\alpha \ln(a+1)+\beta}}$$

where a is the number of active friends, α the social correlation coefficient and β a constant to explain the innate bias for activation. Suppose at one time point t, $Y_{a,t}$ users with a active friends become active, and $N_{a,t}$ users who also have a active friends yet stay inactive at time t. Thus, the likelihood at time t can be expressed as follows:

$$\Pi_t \Pi_a p(a)^{Y_{a,t}} (1 - p(a))^{N_{a,t}}$$

Given the user activity log, we can compute a correlation coefficient α to maximize the above likelihood.

The key idea of the shuffle test is that if influence does not play a role, the timing of activation should be independent of the timing of other actors. Thus, even if we randomly shuffle the timestamps of user activities, we should obtain a similar α value. It is shown that when influence is not a factor, the new estimate of α will be close to its expectation even if the timestamps are randomly shuffled. Hence, we have the following shuffle test for influence:

> **Test for Influence:** After we shuffle the timestamps of user activities, if the new estimate of social correlation is significantly different from the estimate based on the user activity log, then there is evidence of influence.

Beside the shuffle test, many different tests for influence are also proposed for different contexts, including the edge-reversal test (Christakis and Fowler, 2007; Anagnostopoulos et al., 2008), matched sample estimation (Aral et al., 2009) and randomization test (La Fond and Neville, 2010). Employing influence tests, some observe influence effect in obesity (Christakis and Fowler, 2007), service adopters in a telecommunication firm (Hill et al., 2006), and article edits in Wikipedia (Crandall et al., 2008), while some argue that there is no strong influence effect for the tagging behavior in Flickr (Anagnostopoulos et al., 2008). La Fond and Neville (2010) observe that online interest groups on Facebook are formed due to various factors. Those groups with significant homophily effect seem to include opportunities for members to meet in person, while those groups with strong political views seem to be more likely the outcome of influence.

CHAPTER 3

Community Detection and Evaluation

Social networks of various kinds demonstrate a strong community effect. Actors in a network tend to form closely-knit groups. The *groups* are also called *communities*, *clusters*, *cohesive subgroups* or *modules* in different contexts. Generally speaking, individuals interact more frequently with members within group than those outside the group. Detecting cohesive groups in a social network (i.e., *community detection*) remains a core problem in social network analysis. Finding out these groups also helps for other related social computing tasks. Many approaches have been proposed in the past. These approaches can be separated into four categories: node-centric, group-centric, network-centric, and hierarchy-centric (Tang and Liu, 2010b). We introduce definitions and present representative community detection approaches in each category.

3.1   NODE-CENTRIC COMMUNITY DETECTION

The community detection methods based on node-centric criteria require *each node* in a group to satisfy certain properties. We discuss these methods according to these criteria.

3.1.1   COMPLETE MUTUALITY

An ideal cohesive subgroup is a *clique*. It is a maximum complete subgraph in which all nodes are adjacent to each other. For example, in the network in Figure 1.1, there is a clique of 4 nodes, {5, 6, 7, 8}. Typically, cliques of larger sizes are of much more interest. However, the search for the maximum cliques in a graph is an NP-hard problem.

   One brute-force approach is to traverse all nodes in a network. For each node, check whether there is any clique of a specified size that contains the node. Suppose we now look at node v_ℓ. We can maintain a queue of cliques. It is initialized with a clique of one single node $\{v_\ell\}$. Then we perform the following:

- Pop a clique from the queue, say, a clique B_k of size k. Let v_i denote the last added node into B_k.

- For each of v_i's neighbor v_j (to remove duplicates, we may look at only those nodes whose index is larger than v_i), form a new candidate set $B_{k+1} = B_k \cup \{v_j\}$.

- Validate whether B_{k+1} is a clique by checking whether v_j is adjacent to all nodes in B_k. Add to the queue if B_{k+1} is a clique.

Take the network in Figure 1.1 as an example. Suppose we start from node $B_1 = \{4\}$. For each of its friends with a larger index, we obtain a clique of size 2. Thus, we have $\{4, 5\}$ and $\{4, 6\}$ added into the queue. Now suppose we pop $B_2 = \{4, 5\}$ from the queue. Its last added element is node 5. We can expand the set following node 5's connections and generate three candidate sets: $\{4, 5, 6\}$, $\{4, 5, 7\}$ and $\{4, 5, 8\}$. Among them, only $\{4, 5, 6\}$ is a clique as node 6 is connected both nodes 4 and 5. Thus, $\{4, 5, 6\}$ is appended to the queue for further expansion for larger cliques.

The exhaustive search above works for small-scale networks, but it becomes impractical for large-scale networks. If the goal is to find out a maximum clique, then a strategy is to effectively prune those nodes and edges that are unlikely to be contained in the maximum clique. *For a clique of size k, each node in the clique should maintain at least degree k − 1.* Hence, those nodes with degree less than $k - 1$ cannot be included in the maximum clique, thus can be pruned. We can recursively apply the pruning procedure below to a given network:

- A sub-network is sampled from the given network. A clique in the sub-network can be found in a greedy manner, e.g., expanding a clique by adding an adjacent node with the highest degree.

- The maximum clique found on the sub-network (say, it contains k nodes) serves as the lower bound for pruning. That is, the maximum clique in the original network should contain at least k members. Hence, in order to find a clique of size larger than k, the nodes with degree less than or equal to $k - 1$, in conjunction with their connections can be removed from future consideration. As social media networks follow a power law distribution for node degrees, i.e., the majority of nodes have a low degree, this pruning strategy can reduce the network size significantly.

This process is repeated until the original network is shrunk into a reasonable size and the maximum clique can either be identified directly, or have already been identified in one of the sub-networks. A similar pruning strategy is discussed for directed networks as well (Kumar et al., 1999).

Suppose we randomly sample a sub-network from the network in Figure 1.1. It consists of nodes 1 to 6. A maximal clique in the sub-network is of size 3 ($\{1, 2, 3\}$ or $\{1, 3, 4\}$). If there exists a larger clique (i.e., size > 3) in the original network, all the nodes of degree less than or equal to 2 can be removed from consideration. Hence, nodes 9 and 2 can be pruned. Then, the degree of nodes 1 and 3 is reduced to 2, thus they can also be removed. This further leaves node 4 with only 2 connections, which can be removed as well. After this pruning, we obtain a much smaller network of nodes $\{5, 6, 7, 8\}$. And in this pruned network, a clique of size 4 can be identified. It is exactly the maximum clique.

A clique is a very strict definition, and it can rarely be observed in a huge size in real-world social networks. This structure is very unstable as the removal of any edge in it will render it an invalid clique. Practitioners typically use identified cliques as cores or seeds for subsequent expansion for a

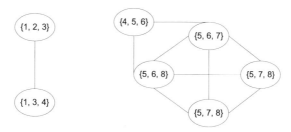

Figure 3.1: A Clique Graph

community. The *clique percolation method* (CPM) is such a scheme to find overlapping communities in networks (Palla et al., 2005). Given a user specified parameter k, it works as follows:

- Find out all cliques of size k in the given network;

- Construct a clique graph. Two cliques are adjacent if they share $k - 1$ nodes;

- Each connected component in the clique graph is a community.

Take the network in Figure 1.1 as an example. For $k = 3$, we can identify all the cliques of size 3 as follows:

$\{1, 2, 3\}$	$\{1, 3, 4\}$	$\{4, 5, 6\}$	$\{5, 6, 7\}$	$\{5, 6, 8\}$	$\{5, 7, 8\}$	$\{6, 7, 8\}$

Then we have the clique graph as in Figure 3.1. Two cliques are connected as long as they share $k - 1$ (2 in our case) nodes. In the clique graph, there are two connected components. The nodes in each component fall into one community. Consequently, we obtain two communities: $\{1, 2, 3, 4\}$ and $\{4, 5, 6, 7, 8\}$. Note that node 4 belongs to both communities. In other words, we obtain two overlapping communities.

The clique percolation method requires the enumeration of all the possible cliques of a fixed size k. This can be computational prohibitive for large-scale social media networks. Other forms of subgraph close to a clique thus are proposed to capture the community structure, which will be discussed next.

3.1.2   REACHABILITY

This type of community considers the reachability among actors. In the extreme case, two nodes can be considered as belonging to one community if there exists a path between the two nodes. Thus each connected component is a community. The components can be efficiently identified in $O(n + m)$ time (Hopcroft and Tarjan, 1973), linear with respect to number of nodes and edges in a network. However, in real-world networks, a giant component tends to form while many others are singletons and minor communities (Kumar et al., 2006). Those minor communities can be identified as connected components. Yet more efforts are required to find communities in the giant component.

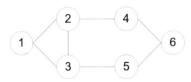

cliques: {1, 2, 3}
2-cliques: {1, 2, 3, 4, 5, }, {2, 3, 4, 5, 6}
2-clubs: {1, 2, 3, 4, }, {1, 2, 3, 5}, {2, 3, 4, 5, 6}

Figure 3.2: An example to show the difference of k-clique and k-club (based on (Wasserman and Faust, 1994))

Conceptually, there should be a short path between any two nodes in a group. Some well-studied structures in social sciences are the following:

- k-*clique* is a maximal subgraph in which the largest geodesic distance between any two nodes is no greater than k. That is,

$$d(v_i, v_j) \leq k \ \ \forall v_i, v_j \in V_s$$

where V_s is the set of nodes in the subgraph. Note that the geodesic distance is defined on the original network. Thus, the geodesic is not necessarily included in the group structure. So a k-clique may have a diameter greater than k. For instance, in Figure 3.2, {1, 2, 3, 4, 5} form a 2-clique. But the geodesic distance between nodes 4 and 5 within the group is 3.

- k-*club* restricts the geodesic distance within the group to be no greater than k. It is a maximal substructure of diameter k. The definition of k-club is more strict than that of k-clique. A k-club is often a subset of a k-clique. In the example in Figure 3.2, The 2-clique structure {1, 2, 3, 4, 5} contains two 2-clubs, {1, 2, 3, 4} and {1, 2, 3, 5}.

There are other definitions of communities such as k-plex, k-core, LS sets, and Lambda sets (Wasserman and Faust, 1994). They are typically studied in traditional social sciences. Solving the k-club problem often requires involved combinatorial optimization (McClosky and Hicks, 2009). It remains a challenge to generalize them to large-scale networks.

3.2   GROUP-CENTRIC COMMUNITY DETECTION

A group-centric criterion considers connections inside a group as whole. It is acceptable to have some nodes in the group to have low connectivity as long as the group overall satisfies certain requirements. One such example is *density-based groups*. A subgraph $G_s(V_s, E_s)$ is γ-dense (also called a *quasi-clique* (Abello et al., 2002)) if

$$\frac{E_s}{V_s(V_s - 1)/2} \geq \gamma. \tag{3.1}$$

Clearly, the quasi-clique becomes a clique when $\gamma = 1$. Note that this density-based group-centric criterion does not guarantee reachability for each node in the group. It allows the degree of a node to vary[1], thus is more suitable for large-scale networks.

However, it is not a trivial task to search for quasi-cliques. Strategies similar to those of finding cliques can be exploited. In Abello et al. (2002), a greedy search followed by pruning is employed to find the maximal γ-dense quasi-clique in a network. The iterative procedure consists of two steps - local search and heuristic pruning.

- *Local search*: Sample a sub-network from the given network and search for a maximal quasi-clique in the sub-network. A greedy approach is to aggressively expand a quasi-clique by encompassing those high-degree neighboring nodes until the density drops below γ. In practice, a randomized search strategy can also be exploited.

- *Heuristic pruning*: If we know a γ-dense quasi-clique of size k, then a heuristic is to prune those "peelable" nodes and their incident edges. A node v is *peelable* if v and its neighbors all have degree less than $k\gamma$ because it is less likely to contribute to a larger quasi-clique by including such a node. We can start from low-degree nodes and recursively remove peelable nodes in the original network.

This process is repeated until the network is reduced to a reasonable size so that a maximal quasi-clique can be found directly. Though the solution returned by the algorithm does not guarantee to be optimal, it works reasonably well in most cases (Abello et al., 2002).

3.3 NETWORK-CENTRIC COMMUNITY DETECTION

Network-centric community detection has to consider the global topology of a network. It aims to *partition* nodes of a network into a number of disjoint sets. Typically, network-centric community detection aims to optimize a criterion defined over a network partition rather than over one group. A group in this case is not defined independently.

3.3.1 VERTEX SIMILARITY

Vertex similarity is defined in terms of the similarity of their social circles, e.g., the number of friends two share in common. A key related concept is *structural equivalence*. Actors v_i and v_j are structurally equivalent, if for any actor v_k that $v_k \neq v_i$ and $v_k \neq v_j$, $e(v_i, v_k) \in E$ iff $e(v_j, v_k) \in E$. In other words, actors v_i and v_j are connecting to exactly the same set of actors in a network. If the interaction is represented as a matrix, then rows (columns) of v_i and v_j are the same except for the diagonal entries. Nodes 1 and 3 in Figure 1.1 are structurally equivalent. So are nodes 5 and 6. A closer examination at its adjacency matrix (Table 1.3) reveals that those structurally equivalent nodes share the same rows (columns). Nodes of the same equivalence class form a community.

---

[1] It removes the need of being connected to at least k other nodes in the same group.

Since structural equivalence is too restrictive for practical use, other relaxed definitions of equivalence such as *automorphic equivalence* and *regular equivalence* are proposed (Hanneman and Riddle, 2005), but no scalable approach exists to find automorphic equivalence or regular equivalence. Alternatively, some simplified similarity measures can be used. They
consider one's connections as features for actors, and they assume actors sharing similar connections tend to reside within the same community. Once a similarity measure is determined, classical
k-means clustering or hierarchical clustering algorithm (Tan et al., 2005) can be applied to find
communities in a network.

Commonly used similarity measures include Jaccard similarity (Gibson et al., 2005) and cosine
similarity (Hopcroft et al., 2003). For two nodes v_i and v_j in a network, the similarity between the
two are defined as

$$Jaccard(v_i, v_j) = \frac{|N_i \cap N_j|}{|N_i \cup N_j|} = \frac{\sum_k A_{ik} A_{jk}}{|N_i| + |N_j| - \sum_k A_{ik} A_{jk}}, \quad (3.2)$$

$$Cosine(v_i, v_j) = \frac{\sum_k A_{ik} A_{jk}}{\sqrt{\sum_s A_{is}^2 \cdot \sum_t A_{jt}^2}} = \frac{|N_i \cap N_j|}{\sqrt{|N_i| \cdot |N_j|}}. \quad (3.3)$$

In the equations, N_i denotes the neighbors of node v_i and $|*|$ the cardinality. Both similarity
measures are within the range between 0 and 1.

For example, in the network in Figure 1.1, $N_4 = \{1, 3, 5, 6\}$, and $N_6 = \{4, 5, 7, 8\}$. Thus, the
similarity between the two nodes are:

$$Jaccard(4, 6) = \frac{|\{5\}|}{|\{1, 3, 4, 5, 6, 7, 8\}|} = \frac{1}{7}, \quad (3.4)$$

$$Cosine(4, 6) = \frac{|\{5\}|}{\sqrt{4 \cdot 4}} = \frac{1}{4}. \quad (3.5)$$

However, based on this definition, the similarity of two adjacent nodes could be 0. For example, the similarity of nodes 7 and 9 is 0 because $N_7 = \{5, 6, 8, 9\}$, $N_9 = \{7\}$, and $N_7 \cap N_9 = \phi$,
even though they are connected. This is reasonable from the perspective of structural equivalence.
However, from the correlation aspect, statistically, two nodes are likely to share some similarity if
they are connected. A modification is to include node v when we compute N_v. In equivalence, the
diagonal entries of the adjacency matrix of a network is set to 1 rather than default 0. In this case,
$N_7 = \{5, 6, 7, 8, 9\}$, $N_9 = \{7, 9\}$. It follows that $N_7 \cap N_9 = \{7, 9\}$:

$$Jaccard(7, 9) = \frac{|\{7, 9\}|}{|\{5, 6, 7, 8, 9\}|} = \frac{2}{5},$$

$$Cosine(7, 9) = \frac{|\{7, 9\}|}{\sqrt{2 \cdot 5}} = \frac{2}{\sqrt{10}}.$$

Normal similarity-based methods have to compute the similarity for each pair of nodes,
totaling $O(n^2)$. It is time-consuming when n is very large. Thus, Gibson et al. (2005) present an

efficient two-level shingling algorithm for fast computation of web communities. Generally speaking, the *shingling* algorithm maps each vector (the connection of actors) into a constant number of "shingles". If two actors are similar, they share many shingles; otherwise, they share few. After initial shingling, each shingle is associated with a group of actors. In a similar vein, the shingling algorithm can be applied to the first-level shingles as well. So similar shingles end up sharing the same meta-shingles. Then all the actors relating to one meta-shingle form one community. This two-level shingling can be efficiently computed even for large-scale networks. Its time complexity is approximately linear to the number of edges.

3.3.2   LATENT SPACE MODELS

A latent space model maps nodes in a network into a low-dimensional Euclidean space such that the proximity between nodes based on network connectivity are kept in the new space (Hoff et al., 2002; Handcock et al., 2007), then the nodes are clustered in the low-dimensional space using methods like k-means (Tan et al., 2005). One representative approach is *multi-dimensional scaling* (MDS) (Borg and Groenen, 2005). Typically, MDS requires the input of a proximity matrix $P \in \mathbb{R}^{n \times n}$, with each entry P_{ij} denoting the distance between a pair of nodes i and j in the network. Let $S \in \mathbb{R}^{n \times \ell}$ denote the coordinates of nodes in the ℓ-dimensional space such that S is column orthogonal. It can be shown (Borg and Groenen, 2005; Sarkar and Moore, 2005) that

$$SS^T \approx -\frac{1}{2}(I - \frac{1}{n}\mathbf{1}\mathbf{1}^T)(P \circ P)(I - \frac{1}{n}\mathbf{1}\mathbf{1}^T) = \widetilde{P}, \tag{3.6}$$

where I is the identity matrix, $\mathbf{1}$ an n-dimensional column vector with each entry being 1, and $\circ$ the element-wise matrix multiplication. It follows that S can be obtained via minimizing the discrepancy between $\widetilde{P}$ and SS^T as follows:

$$\min \|SS^T - \widetilde{P}\|_F^2. \tag{3.7}$$

Suppose V contains the top ℓ eigenvectors of $\widetilde{P}$ with largest eigenvalues, Λ is a diagonal matrix of top ℓ eigenvalues $\Lambda = diag(\lambda_1, \lambda_2, \cdots, \lambda_\ell)$. The optimal S is $S = V \Lambda^{\frac{1}{2}}$. Note that this multi-dimensional scaling corresponds to an eigenvector problem of matrix $\widetilde{P}$. Thus, the classical k-means algorithm can be applied to S to find community partitions.

Take the network in Figure 1.1 as an example. Given the network, the geodesic distance between each pair of nodes is given in the proximity matrix P as in Eq. (3.8). Hence, we can compute the corresponding matrix $\widetilde{P}$ following Eq. (3.6).

$$
P = \begin{bmatrix}
0 & 1 & 1 & 1 & 2 & 2 & 3 & 3 & 4 \\
1 & 0 & 1 & 2 & 3 & 3 & 4 & 4 & 5 \\
1 & 1 & 0 & 1 & 2 & 2 & 3 & 3 & 4 \\
1 & 2 & 1 & 0 & 1 & 1 & 2 & 2 & 3 \\
2 & 3 & 2 & 1 & 0 & 1 & 1 & 1 & 2 \\
2 & 3 & 2 & 1 & 1 & 0 & 1 & 1 & 2 \\
3 & 4 & 3 & 2 & 1 & 1 & 0 & 1 & 1 \\
3 & 4 & 3 & 2 & 1 & 1 & 1 & 0 & 2 \\
4 & 5 & 4 & 3 & 2 & 2 & 1 & 2 & 0
\end{bmatrix},
\tag{3.8}
$$

$$
\widetilde{P} = \begin{bmatrix}
2.46 & 3.96 & 1.96 & 0.85 & -0.65 & -0.65 & -2.21 & -2.04 & -3.65 \\
3.96 & 6.46 & 3.96 & 1.35 & -1.15 & -1.15 & -3.71 & -3.54 & -6.15 \\
1.96 & 3.96 & 2.46 & 0.85 & -0.65 & -0.65 & -2.21 & -2.04 & -3.65 \\
0.85 & 1.35 & 0.85 & 0.23 & -0.27 & -0.27 & -0.82 & -0.65 & -1.27 \\
-0.65 & -1.15 & -0.65 & -0.27 & 0.23 & -0.27 & 0.68 & 0.85 & 1.23 \\
-0.65 & -1.15 & -0.65 & -0.27 & -0.27 & 0.23 & 0.68 & 0.85 & 1.23 \\
-2.21 & -3.71 & -2.21 & -0.82 & 0.68 & 0.68 & 2.12 & 1.79 & 3.68 \\
-2.04 & -3.54 & -2.04 & -0.65 & 0.85 & 0.85 & 1.79 & 2.46 & 2.35 \\
-3.65 & -6.15 & -3.65 & -1.27 & 1.23 & 1.23 & 3.68 & 2.35 & 6.23
\end{bmatrix}.
$$

Suppose we want to map the original network into a 2-dimensional space; we obtain V, Λ, and S as follows:

$$
V = \begin{bmatrix}
-0.33 & 0.05 \\
-0.55 & 0.14 \\
-0.33 & 0.05 \\
-0.11 & -0.01 \\
0.10 & -0.06 \\
0.10 & -0.06 \\
0.32 & 0.11 \\
0.28 & -0.79 \\
0.52 & 0.58
\end{bmatrix}, \quad
\Lambda = \begin{bmatrix}
21.56 & 0 \\
0 & 1.46
\end{bmatrix}, \quad
S = V\Lambda^{1/2} = \begin{bmatrix}
-1.51 & 0.06 \\
-2.56 & 0.17 \\
-1.51 & 0.06 \\
-0.53 & -0.01 \\
0.47 & -0.08 \\
0.47 & -0.08 \\
1.47 & 0.14 \\
1.29 & -0.95 \\
2.42 & 0.70
\end{bmatrix}.
$$

The network can be visualized in the 2-dimensional space in Figure 3.3. Because nodes 1 and 3 are structurally equivalent, they are mapped into the same position in the latent space. So are nodes 5 and 6. k-means can be applied to S in order to obtain disjoint partitions of the network. At the end, we obtain two clusters $\{1, 2, 3, 4\}$, $\{5, 6, 7, 8, 9\}$, which can be represented as a partition

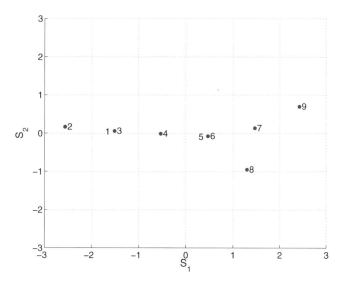

Figure 3.3: Network in the Latent Space

matrix below:

$$H = \begin{bmatrix} 1 & 0 \\ 1 & 0 \\ 1 & 0 \\ 1 & 0 \\ 0 & 1 \\ 0 & 1 \\ 0 & 1 \\ 0 & 1 \\ 0 & 1 \end{bmatrix}.$$

3.3.3   BLOCK MODEL APPROXIMATION

Block models approximate a given network by a block structure. The basic idea can be visualized in Tables 3.1 and 3.2. The adjacency matrix of the network in Figure 1.1 is shown in Table 3.1. We highlight those entries that indicate an edge between two nodes. The adjacency matrix can be approximated by a block structure as shown in Table 3.2. Each block represents one community. Therefore, we approximate a given adjacency matrix A as follows:

$$A \approx S\Sigma S^T, \tag{3.9}$$

where $S \in \{0, 1\}^{n \times k}$ is the block indicator matrix with $S_{ij} = 1$ if node i belongs to the j-th block, Σ a $k \times k$ matrix indicating the block (group) interaction density, and k the number of blocks. A

Table 3.1: Adjacency Matrix

-	1	1	1	0	0	0	0	0
1	-	1	0	0	0	0	0	0
1	1	-	1	0	0	0	0	0
1	0	1	-	1	1	0	0	0
0	0	0	1	-	1	1	1	0
0	0	0	1	1	-	1	1	0
0	0	0	0	1	1	-	1	1
0	0	0	0	1	1	1	-	0
0	0	0	0	0	0	1	0	-

Table 3.2: Ideal Block Structure

1	1	1	1	0	0	0	0	0
1	1	1	1	0	0	0	0	0
1	1	1	1	0	0	0	0	0
1	1	1	1	0	0	0	0	0
0	0	0	0	1	1	1	1	1
0	0	0	0	1	1	1	1	1
0	0	0	0	1	1	1	1	1
0	0	0	0	1	1	1	1	1
0	0	0	0	1	1	1	1	1

natural objective is to minimize the following:

$$\min \|A - S \Sigma S^T\|_F^2. \tag{3.10}$$

The discreteness of S makes the problem NP-hard. We can relax S to be continuous but satisfy certain orthogonal constraints, i.e., $S^T S = I_k$, then the optimal S corresponds to the top k eigenvectors of A with maximum eigenvalues. Similar to the latent space model, k-means clustering can be applied to S to recover the community partition H.

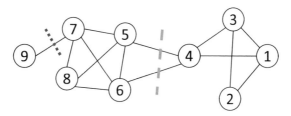

Figure 3.4: Two Different Cuts of the Toy Network in Figure 1.1

For the network in Figure 1.1, the top two eigenvectors of the adjacency matrix are

$$S = \begin{bmatrix} 0.20 & -0.52 \\ 0.11 & -0.43 \\ 0.20 & -0.52 \\ 0.38 & -0.30 \\ 0.47 & 0.15 \\ 0.47 & 0.15 \\ 0.41 & 0.28 \\ 0.38 & 0.24 \\ 0.12 & 0.11 \end{bmatrix}, \Sigma = \begin{bmatrix} 3.5 & 0 \\ 0 & 2.4 \end{bmatrix}.$$

As indicated by the sign of the second column of S, nodes $\{1, 2, 3, 4\}$ form a community, and $\{5, 6, 7, 8, 9\}$ is another community, which can be obtained by a k-means clustering applied to S.

3.3.4   SPECTRAL CLUSTERING

Spectral clustering (Luxburg, 2007) is derived from the problem of graph partition. Graph partition aims to find out a partition such that the cut (the total number of edges between two disjoint sets of nodes) is minimized. For instance, the green cut (thick dashed line) between two sets of nodes $\{1, 2, 3, 4\}$ and $\{5, 6, 7, 8, 9\}$ in Figure 3.4 is 2 as there are two edges $e(4, 5)$ and $e(4, 6)$. Intuitively, if two communities are well separated, the cut between them should be small. Hence, a community detection problem can be reduced to finding the minimum cut in a network. This *minimum cut* problem can be solved efficiently. It, however, often returns imbalanced communities, with one being trivial or a singleton, i.e., a community consisting of only one node. In the network in Figure 3.4, for example, the minimum cut is 1, between $\{9\}$ and $\{1, 2, 3, 4, 5, 6, 7, 8\}$.

Therefore, the objective function is modified so that the group sizes of communities are considered. Two commonly used variants are *ratio cut* and *normalized cut*. Let $\pi = (C_1, C_2, \cdots, C_k)$ be a graph partition such that $C_i \cap C_j = \phi$ and $\cup_{i=1}^{k} C_i = V$. The ratio cut and the normalized cut are defined as:

$$\text{Ratio Cut}(\pi) = \frac{1}{k} \sum_{i=1}^{k} \frac{cut(C_i, \bar{C}_i)}{|C_i|}, \tag{3.11}$$

$$\text{Normalized Cut}(\pi) = \frac{1}{k} \sum_{i=1}^{k} \frac{cut(C_i, \bar{C}_i)}{vol(C_i)}. \tag{3.12}$$

where $\bar{C}_i$ is the complement of C_i, and $vol(C_i) = \sum_{v \in C_i} d_v$. Both objectives attempt to minimize the number of edges between communities, yet avoid the bias of trivial-size communities like singletons.

Suppose we partition the network in Figure 1.1 into two communities, with $C_1 = \{9\}$ and $C_2 = \{1, 2, 3, 4, 5, 6, 7, 8\}$. Let this partition be denoted as π_1. It follows that $cut(C_1, \bar{C}_1) = 1$, $|C_1| = 1$, $|C_2| = 8$, $vol(C_1) = 1$, and $vol(C_2) = 27$. Consequently,

$$\text{Ratio Cut}(\pi_1) = \frac{1}{2} \left(\frac{1}{1} + \frac{1}{8} \right) = 9/16 = 0.56,$$

$$\text{Normalized Cut}(\pi_1) = \frac{1}{2} \left(\frac{1}{1} + \frac{1}{27} \right) = 14/27 = 0.52.$$

Now for another more balanced partition π_2 with $C_1 = \{1, 2, 3, 4\}$, and $C_2 = \{5, 6, 7, 8, 9\}$, we have

$$\text{Ratio Cut}(\pi_2) = \frac{1}{2} \left(\frac{2}{4} + \frac{2}{5} \right) = 9/20 = 0.45 < \text{Ratio Cut}(\pi_1),$$

$$\text{Normalized Cut}(\pi_2) = \frac{1}{2} \left(\frac{2}{12} + \frac{2}{16} \right) = 7/48 = 0.15 < \text{Normalized Cut}(\pi_1).$$

Though the cut of partition π_1 is smaller, partition π_2 is preferable based on the ratio cut or the normalized cut.

Nevertheless, finding the minimum ratio cut or normalized cut is NP-hard. An approximation is to use spectral clustering. Both ratio cut and normalized cut can be formulated as a min-trace problem like below

$$\min_{S \in \{0,1\}^{n \times k}} Tr(S^T \tilde{L} S), \tag{3.13}$$

with the (normalized) graph Laplacian $\tilde{L}$ defined as follows:

$$\tilde{L} = \begin{cases} D - A & \text{(Graph Laplacian for Ratio Cut)} \\ I - D^{-1/2} A D^{-1/2} & \text{(Normalized Graph Laplacian for Normalized Cut)} \end{cases} \tag{3.14}$$

with $D = diag(d_1, d_2, \cdots, d_n)$. Akin to block model approximation, we solve the following spectral clustering problem based on a relaxation to S (Luxburg, 2007).

$$\min_{S} Tr(S^T \tilde{L} S) \quad s.t. \ S^T S = I_k \tag{3.15}$$

Then, S corresponds to the top eigenvectors of $\widetilde{L}$ with the smallest eigenvalues. For the network in Figure 1.1,

$$D = diag(3, 2, 3, 4, 4, 4, 4, 3, 1);$$

and the graph Laplacian is

$$\widetilde{L} = D - A = \begin{bmatrix} 3 & -1 & -1 & -1 & 0 & 0 & 0 & 0 & 0 \\ -1 & 2 & -1 & 0 & 0 & 0 & 0 & 0 & 0 \\ -1 & -1 & 3 & -1 & 0 & 0 & 0 & 0 & 0 \\ -1 & 0 & -1 & 4 & -1 & -1 & 0 & 0 & 0 \\ 0 & 0 & 0 & -1 & 4 & -1 & -1 & -1 & 0 \\ 0 & 0 & 0 & -1 & -1 & 4 & -1 & -1 & 0 \\ 0 & 0 & 0 & 0 & -1 & -1 & 4 & -1 & -1 \\ 0 & 0 & 0 & 0 & -1 & -1 & -1 & 3 & 0 \\ 0 & 0 & 0 & 0 & 0 & 0 & -1 & 0 & 1 \end{bmatrix},$$

with its two smallest eigenvectors being

$$S = \begin{bmatrix} 0.33 & -0.38 \\ 0.33 & -0.48 \\ 0.33 & -0.38 \\ 0.33 & -0.12 \\ 0.33 & 0.16 \\ 0.33 & 0.16 \\ 0.33 & 0.30 \\ 0.33 & 0.24 \\ 0.33 & 0.51 \end{bmatrix}.$$

Typically, the first eigenvector does not contain any community information. For the example above, all the nodes are assigned with the same value, meaning that all reside in the same community. Thus, the first eigenvector is often discarded. In order to find out k communities, $k - 1$ smallest eigenvectors (except the first one) are used to feed into k-means for clustering. In our example, the second column of S, as indicated by the sign, tells us that the network can be divided into two groups $\{1, 2, 3, 4\}$ and $\{5, 6, 7, 8, 9\}$.

3.3.5 MODULARITY MAXIMIZATION

Modularity (Newman, 2006a) is proposed specifically to measure the strength of a community partition for real-world networks by taking into account the degree distribution of nodes. Given a network of n nodes and m edges, the *expected number of edges* between nodes v_i and v_j is $d_i d_j / 2m$, where d_i and d_j are the degrees of node v_i and v_j, respectively. Considering one edge from node v_i connecting to all nodes in the network randomly, it lands at node v_j with probability $d_j / 2m$. As there are d_i such edges, the expected number of connections between the two are $d_i d_j / 2m$. For example, the network in Figure 1.1 has 9 nodes and 14 edges. The expected number of edges between nodes 1 and 2 is $3 \times 2/(2 \times 14) = 3/14$.

So $A_{ij} - d_i d_j / 2m$ measures how far the true network interaction between nodes i and j (A_{ij}) deviates from the expected random connections. Given a group of nodes C, the strength of community effect is defined as

$$\sum_{i \in C, j \in C} A_{ij} - d_i d_j / 2m.$$

If a network is partitioned into k groups, the overall community effect can be summed up as follows:

$$\sum_{\ell=1}^{k} \sum_{i \in C_\ell, j \in C_\ell} \left(A_{ij} - d_i d_j / 2m \right).$$

Modularity is defined as

$$Q = \frac{1}{2m} \sum_{\ell=1}^{k} \sum_{i \in C_\ell, j \in C_\ell} \left(A_{ij} - d_i d_j / 2m \right). \tag{3.16}$$

where the coefficient $1/2m$ is introduced to normalize the value between -1 and 1. Modularity calibrates the quality of community partitions thus can be used as an objective measure to maximize.

We can define a modularity matrix B as $B_{ij} = A_{ij} - d_i d_j / 2m$. Equivalently,

$$B = A - \mathbf{d}\mathbf{d}^T / 2m. \tag{3.17}$$

where $\mathbf{d} \in R^{n \times 1}$ is a vector of each node's degree. Let $S \in \{0, 1\}^{n \times k}$ be a community indicator matrix with $S_{i\ell} = 1$ if node i belongs to community C_ℓ, and s_ℓ the ℓ-th column of S. Modularity can be reformulated as

$$Q = \frac{1}{2m} \sum_{\ell=1}^{k} s_\ell B s_\ell = \frac{1}{2m} Tr(S^T B S). \tag{3.18}$$

With a spectral relaxation to allow S to be continuous, the optimal S can be computed as the top k eigenvectors of the modularity matrix B (Newman, 2006b) with the maximum eigenvalues. For example, the modularity matrix of the toy network is

$$B = \begin{bmatrix}
-0.32 & 0.79 & 0.68 & 0.57 & -0.43 & -0.43 & -0.43 & -0.32 & -0.11 \\
0.79 & -0.14 & 0.79 & -0.29 & -0.29 & -0.29 & -0.29 & -0.21 & -0.07 \\
0.68 & 0.79 & -0.32 & 0.57 & -0.43 & -0.43 & -0.43 & -0.32 & -0.11 \\
0.57 & -0.29 & 0.57 & -0.57 & 0.43 & 0.43 & -0.57 & -0.43 & -0.14 \\
-0.43 & -0.29 & -0.43 & 0.43 & -0.57 & 0.43 & 0.43 & 0.57 & -0.14 \\
-0.43 & -0.29 & -0.43 & 0.43 & 0.43 & -0.57 & 0.43 & 0.57 & -0.14 \\
-0.43 & -0.29 & -0.43 & -0.57 & 0.43 & 0.43 & -0.57 & 0.57 & 0.86 \\
-0.32 & -0.21 & -0.32 & -0.43 & 0.57 & 0.57 & 0.57 & -0.32 & -0.11 \\
-0.11 & -0.07 & -0.11 & -0.14 & -0.14 & -0.14 & 0.86 & -0.11 & -0.04
\end{bmatrix}$$

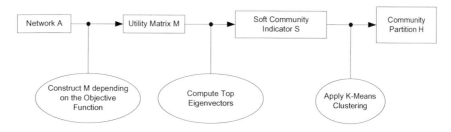

Figure 3.5: A Unified Process for Some Representative Community Detection Methods

Its top two maximum eigenvectors are

$$
S = \begin{bmatrix}
0.44 & -0.00 \\
0.38 & 0.23 \\
0.44 & -0.00 \\
0.17 & -0.48 \\
-0.29 & -0.32 \\
-0.29 & -0.32 \\
-0.38 & 0.34 \\
-0.34 & -0.08 \\
-0.14 & 0.63
\end{bmatrix}.
$$

Note how the partition information is expressed by the first column of S.

3.3.6   A UNIFIED PROCESS

The above four representative community detection methods - latent space models, block model approximation, spectral clustering, and modularity maximization - can be unified in a process as in Figure 3.5. The process is composed of four components with three intermediate steps. Given a network, a utility matrix is constructed. Depending on the objective function, we can construct different utility matrices:

$$
\text{Utility Matrix } M = \begin{cases}
\widetilde{P} \text{ in Eq. (3.6)} & \text{(latent space models)} \\
A \text{ in Eq. (3.9)} & \text{(block model approximation)} \\
\widetilde{L} \text{ in Eq. (3.14)} & \text{(spectral clustering)} \\
B \text{ in Eq. (3.17)} & \text{(modularity maximization)}
\end{cases}
\tag{3.19}
$$

After obtaining the utility matrix, we obtain the *soft community indicator* S that consists of the top eigenvectors with the largest (or smallest subject to formulation) eigenvalues. The selected eigenvectors capture the prominent interaction patterns, representing approximate community partitions. This step can also be considered as a de-noising process since we only keep those top eigenvectors that are indicative of community structures. To recover the discrete partition H, a k-means clustering algorithm is applied. Note that all the aforementioned community detection methods differ subtly by constructing different utility matrices.

The community detection methods presented above, except the latent space model, are normally applicable to medium-size networks (say, $100,000$ nodes). The latent space model requires an input of a proximity matrix of the geodesic distance of any pair of nodes, which costs $O(n^3)$ to compute the pairwise geodesic distances. Moreover, the utility matrix of the latent space model is neither sparse nor structured, incurring $O(n^3)$ time to compute its eigenvectors. This high computational cost hinders its application to real-world large-scale networks. By contrast, block models, spectral clustering and modularity maximization are typically much faster[2].

3.4   HIERARCHY-CENTRIC COMMUNITY DETECTION

Another line of community detection research is to build a hierarchical structure of communities based on network topology. This facilitates the examination of communities at different granularity. There are mainly two types of hierarchical clustering: divisive, and agglomerative.

3.4.1   DIVISIVE HIERARCHICAL CLUSTERING

Divisive clustering first partitions the nodes into several disjoint sets. Then each set is further divided into smaller ones until each set contains only a small number of (say, only one) actors. The key here is how to split a network into several parts. Some partition methods such as block models, spectral clustering, and latent space models can be applied recursively to divide a community into smaller sets.

One particular divisive clustering algorithm receiving much attention is to recursively remove the "weakest" tie in a network until the network is separated into two or more components. The general principle is as follows:

- At each iteration, find out the edge with least strength. This kind of edge is most likely to be a tie connecting two communities.

- Remove the edge and then update the strength of links.

- Once a network is decomposed into two connected components, each component is considered a community. The iterative process above can be applied to each community to find sub-communities.

Newman and Girvan (2004) proposes to find the weak ties based on *edge betweenness*. Edge betweenness is highly related to the betweenness centrality discussed in Section 2.1. Edge betweenness is defined to be the number of shortest paths that pass along one edge (Brandes, 2001). If

---

[2]The utility matrix of modularity maximization is dense, but it is a sparse matrix plus a low rank update as in Eq. (3.17). This structure can be exploited for fast eigenvector computation (Newman, 2006b; Tang et al., 2009).

Table 3.3: Edge Betweenness

	1	2	3	4	5	6	7	8	9
1	0	4	1	9	0	0	0	0	0
2	4	0	4	0	0	0	0	0	0
3	1	4	0	9	0	0	0	0	0
4	9	0	9	0	10	10	0	0	0
5	0	0	0	10	0	1	6	3	0
6	0	0	0	10	1	0	6	3	0
7	0	0	0	0	6	6	0	2	8
8	0	0	0	0	3	3	2	0	0
9	0	0	0	0	0	0	8	0	0

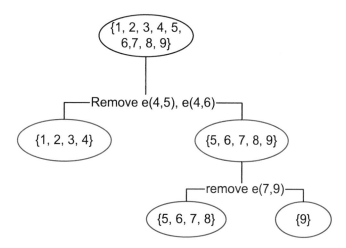

Figure 3.6: The Process of the Newman-Girvan Algorithm Applied to the Toy Network

two communities are joined by only a few cross-group edges, then all paths through the network from nodes in one community to the other community have to pass along one of these edges. Edge betweenness is a measure to count how many shortest paths between pair of nodes pass along the edge, and this number is expected to be large for those between-group edges. The Newman-Girvan algorithm suggests progressively removing edges with the highest betweenness. It will gradually disconnect the network, naturally leading to a hierarchical structure.

The edge betweenness of the network in Figure 1.1 is given in Table 3.3. For instance, the betweenness of $e(1, 2)$ is 4. Since all shortest paths from node 2 to any node in $\{4, 5, 6, 7, 8, 9\}$ has either to pass $e(1, 2)$ or $e(1, 3)$, leading to a weight of $6 \times 1/2 = 3$ for $e(1, 2)$. Meanwhile, $e(1, 2)$ is the shortest path between nodes 1 and 2. Hence, the betweenness of $e(1, 2)$ is $3 + 1 = 4$. An algorithm to compute the edge betweenness is included in Appendix B.

As seen in the table, both edges $e(4, 5)$ and $e(4, 6)$ have highest edge betweenness. Suppose we randomly remove one (say $e(4, 5)$). Then in the resultant network, the edge with the highest betweenness is $e(4, 6)$ (with betweenness being 20). After removing the edge $e(4, 6)$, the network is decomposed into two communities. At this point, $e(7, 9)$ becomes the edge with the highest betweenness. Its removal results in two new communities {5, 6, 7, 8} and {9}. Then a similar procedure can be applied to each community to further divide them into smaller ones. The overall process of the first few steps is shown in Figure 3.6.

However, this divisive hierarchical clustering based on edge betweenness presses hard for computation. As we have discussed in Section 2.1, betweenness for nodes or edges takes $O(nm)$ time (Brandes, 2001). Moreover, each removal of an edge will lead to the recomputation of betweenness for all edges within the same connected component. Its high computational cost hinders its application to large-scale networks.

3.4.2   AGGLOMERATIVE HIERARCHICAL CLUSTERING

Agglomerative clustering begins with base communities and merges them successively into larger communities following certain criterion. One such criterion is modularity (Clauset et al., 2004). Two communities are merged if doing so results in the largest increase of overall modularity. We can start from treating each node as a separate base community and merge communities. The merge continues until no merge can be found to improve the modularity. Figure 3.7 shows the resultant dendrogram based on agglomerative hierarchical clustering applied to the network in Figure 1.1. Nodes 7 and 9 are merged first, and then 1 and 2, and so on. Finally, we obtain two communities at the top {1, 2, 3, 4} and {5, 6, 7, 8, 9}.

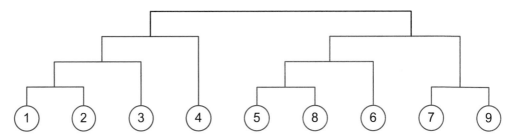

Figure 3.7: Dendrogram according to Agglomerative Clustering based on Modularity

It is noticed that this algorithm incurs many imbalanced merges (i.e., a large community merges with a tiny community, such as the merge of node 4 with {1, 2, 3}), resulting in a high computational cost (Wakita and Tsurumi, 2007). Hence, the merge criterion is modified accordingly by considering the size of communities. In the new scheme, communities of comparable sizes are joined first, leading to a more balanced hierarchical structure of communities and to improved efficiency. A state-of-the-art for such hierarchical clustering is the Louvain method (Blondel et al., 2008). It starts from each node as a base community and determines which of its neighbor should be merged to based on the modularity criterion. After the first scan, some base communities are merged into one. Then, each community's degree and connection information is aggregated and treated as a single node for further merge. This multi-level approach is extremely efficient to handle large-scale networks.

So far, we have discussed some general ideas and different community detection approaches. We can hardly conclude which one is the best. It highly depends on the task at hand and available network data. Other things being equal, we need to resort to evaluation for comparing different community detection methods.

3.5   COMMUNITY EVALUATION

Part of the reason that there are so many assorted definitions and methods, is that there is no clear ground truth information about a community structure in a real world network. Therefore, different community detection methods are developed from various applications of specific needs. We now depict strategies commonly adopted to evaluate identified communities in order to facilitate the comparison of different community detection methods. Depending on available network information, one can take different strategies for comparison:

- Groups with self-consistent definitions. Some groups like cliques, k-cliques, k-clubs, k-plexes and k-cores can be examined immediately once a community is identified. We can simply check whether the extracted communities satisfy the definition.

- Networks with ground truth. That is, the community membership for each actor is known. This is an ideal case. This scenario hardly presents itself in real-world large-scale networks. It usually occurs for evaluation on synthetic networks generated based on predefined community structures (e.g., (Tang et al., 2008)), or some well-studied tiny networks like Zachary's karate club with 34 members (Newman, 2006b). To compare the ground truth with identified community structures, visualization can be intuitive and straightforward (Newman, 2006b). If the number of communities is small (say 2 or 3 communities), it is easy to determine a one-to-one mapping between the identified communities and the ground truth. So conventional classification measures (Tan et al., 2005) such as accuracy, F1-measure can be also used. In Figure 3.8, e.g., one can say one node 2 is wrongly assigned.

  However, when there are many communities, it may not be clear what a correct mapping of communities from the ground truth to a clustering result. In Figure 3.8, the ground truth has two communities whereas the clustering result consists of three. Both communities {1, 3} and {2} map to the community {1, 2, 3} in the ground truth. Hence, some measure that can average all the possible mappings should be considered. Normalized mutual information (NMI) is a commonly used one (Strehl and Ghosh, 2003).

  Before we introduce NMI, we briefly review some information-theoretic concepts. In information theory, the information contained in a distribution is called *entropy*, which is defined

Figure 3.8: Comparing Ground Truth with Clustering Result. Each number denotes a node, and each circle or block denotes a community.

below:

$$H(X) = -\sum_{x \in X} p(x) \log p(x). \tag{3.20}$$

Mutual information calibrates the shared information between two distributions:

$$I(X; Y) = \sum_{y \in Y} \sum_{x \in X} p(x, y) \log \left(\frac{p(x, y)}{p_1(x) p_2(y)} \right). \tag{3.21}$$

Since $I(X; Y) \leq H(X)$ and $I(X; Y) \leq H(Y)$, a *normalized mutual information* (NMI) between two variables X and Y is

$$NMI(X; Y) = \frac{I(X; Y)}{\sqrt{H(X)H(Y)}}. \tag{3.22}$$

We can consider a partition as a probability distribution of one node falling into one community. Let π^a, π^b denote two different partitions of communities. $n_{h,\ell}$, n_h^a, n_ℓ^b are, respectively, the number of actors simultaneously belonging to the h-th community of π^a and ℓ-th community of π^b, the number of actors in the h-th community of partition π^a, and the number of actors in the ℓ-th community of partition π^b. Thus,

$$H(\pi^a) = \sum_{h}^{k^{(a)}} \frac{n_h^a}{n} \log \frac{n_h^a}{n},$$

$$H(\pi^b) = \sum_{\ell}^{k^{(b)}} \frac{n_\ell^b}{n} \log \frac{n_\ell^b}{n},$$

$$I(\pi^a; \pi^b) = \sum_{h} \sum_{\ell} \frac{n_{h,\ell}}{n} \log \left(\frac{\frac{n_{h,\ell}}{n}}{\frac{n_h^a}{n} \frac{n_\ell^b}{n}} \right).$$

$n = 6$		n_h^a		n_l^b	$n_{h,l}$	l=1	l=2	l=3
$k^{(a)} = 2$	h=1	3	l=1	2	h=1	2	1	0
$k^{(b)} = 3$	h=2	3	l=2	1	h=2	0	0	3
			l=3	3				

Figure 3.9: Computation of NMI to compare two clusterings in Figure 3.8.

In the formula, $\frac{n_{h,\ell}}{n}$, in essence, estimates the probability of the mapping from community h in $\pi^{(a)}$ to community ℓ in $\pi^{(b)}$. Consequently,

$$
NMI(\pi^a; \pi^b) = \frac{\sum_{h=1}^{k^{(a)}} \sum_{\ell=1}^{k^{(b)}} n_{h,\ell} \log\left(\frac{n \cdot n_{h,l}}{n_h^{(a)} \cdot n_\ell^{(b)}}\right)}{\sqrt{\left(\sum_{h=1}^{k^{(a)}} n_h^{(a)} \log \frac{n_h^a}{n}\right) \left(\sum_{\ell=1}^{k^{(b)}} n_\ell^{(b)} \log \frac{n_\ell^b}{n}\right)}}.
\tag{3.23}
$$

NMI is a measure between 0 and 1. It equals 1 when π^a and π^b are the same.

As for the example in Figure 3.8, the partitions can be rewritten in another form as follows:

$$
\begin{aligned}
\pi^a &= [1, 1, 1, 2, 2, 2] \quad \text{(ground truth)} \\
\pi^b &= [1, 2, 1, 3, 3, 3] \quad \text{(clustering result)}
\end{aligned}
$$

The network has 6 nodes, with each assigned to one community. Here, the numbers are the community ids of each clustering. The corresponding quantity of each term in Eq. (3.23) is listed in Figure 3.9. The resultant NMI following Eq. (3.23) is 0.83.

Another way is to consider all the possible pairs of nodes and check whether they reside in the same community. It is considered an error if two nodes of the same community are assigned to different communities, or two nodes of different communities are assigned to the same community. Let $C(v_i)$ denote the community of node v_i. We can construct a contingency table below: a, b, c and d are frequencies of each case. For instance, a is the frequency that

		Ground Truth	
		$C(v_i) = C(v_j)$	$C(v_i) \neq C(v_j)$
Clustering	$C(v_i) = C(v_j)$	a	b
Result	$C(v_i) \neq C(v_j)$	c	d

two nodes are assigned into the same community in the ground truth as well in the clustering result. It is noticed that the total sum of frequencies is the number of all possible pairs of nodes in a network, i.e., $a + b + c + d = n(n-1)/2$. Based on the frequencies, the accuracy of clustering can be computed as

$$accuracy = \frac{a+d}{a+b+c+d} = \frac{a+d}{n(n-1)/2}.$$

Take Figure 3.8 as an example. We have $a = 4$. Specifically, $\{1, 3\},\{4, 5\}, \{4, 6\}, \{5, 6\}$ are assigned into the same community in the ground truth and clustering result. Any pair between $\{1, 2, 3\}$ and $\{4, 5, 6\}$ are being assigned to different communities, thus $d = 9$. Consequently, the accuracy of the clustering result is $(4 + 9)/(6 \times 5/2) = 13/15$.

- Networks with semantics. Some networks come with semantic or attribute information of nodes and connections. In this case, the identified communities can be verified by human subjects to check whether it is consistent with the semantics, for instance, whether the community identified in the Web is coherent to a shared topic (Flake et al., 2000; Clauset et al., 2004), and whether the clustering of coauthorship network captures the research interests of individuals. This evaluation approach is applicable when the community is reasonably small. Otherwise, selecting the top-ranking actors as representatives of a community is a commonly used approach. Since this approach is qualitative, it can hardly be applied to all communities in a large network, but it is quite helpful for understanding and interpretation of community patterns. For example, tag clouds of representative nodes in two communities in blogosphere are shown in Figure 3.10 (Tang, 2010). Though these two communities are extracted based on network topology, they both capture certain semantic meanings. The first community is about *animals*, and the other one is about *health*.

Figure 3.10: Tag Clouds of Extracted Communities based on (Tang, 2010)

- Networks without ground truth or semantic information. This is the most common situation, yet it requires objective evaluation most. Normally, one resorts to some quantitative measure for network validation. That is, the quality measure Q is a function of a partition π and a network A. We can use a similar procedure as cross validation in classification for validation. It extracts communities from a (training) network and then compares them with those of the same network (e.g., constructed from a different date) or another related network based on a different type of interaction.

In order to quantify the quality of extracted community structure, a common measure being used is modularity (Newman, 2006a). Once we have a network partition, we can compute its modularity with respect to one network. The method with higher modularity wins. Another comparable approach is to use the identified community as a base for link prediction, i.e., two actors are connected if they belong to the same community. Then, the predicted network is compared with the true network, and the deviation is used to calibrate the community structure. Since social media networks demonstrate strong community effect, a better community structure should predict the connections between actors more accurately. This is basically checking how far the true network deviates from a block model based on the identified communities.

Community detection is still an active and evolving field. We present some widely used community detection and evaluation strategies. In (Fortunato, 2010), one can find a comprehensive survey. Social media, however, often presents more than just a single friendship network. It might involve heterogeneous types of entities and interactions. In the next chapter, we will discuss further how to integrate different kinds of interaction information together to find robust communities in social media.

CHAPTER 4

Communities in Heterogeneous Networks

With social media, people can interact with each other more conveniently than ever. Examining activities of users, we can observe different interaction networks between the same set of actors. Take YouTube as an example. A user may become a friend of another user's; he can also "subscribe" to another user. The existence of different types of interactions suggests heterogeneous interactions in one social network. Meanwhile, in some social networking sites, entities other than human beings can also be involved. For instance, in YouTube, a *user* can upload a *video* and another user can *tag* it. In other words, users, videos, and tags are weaved into the same network. The "actors" in the network are not at all homogeneous. Networks involving heterogeneous types of interactions or actors are referred as *heterogeneous networks*. This chapter discusses how we might extend methods presented in the previous chapter to handle this heterogeneity.

4.1   HETEROGENEOUS NETWORKS

Social media networks are often heterogeneous, having heterogeneous interactions or heterogeneous nodes. Heterogeneous networks are thus categorized into *multi-dimensional networks* and *multi-mode networks* as defined below:

- *Multi-Dimensional Networks*. A multi-dimensional network has multiple types of interactions between the same set of users. Each dimension of the network represents one type of activity between users. For instance, at popular content sharing sites like Flickr and YouTube, as in Figure 4.1, a user can connect to his friends through email invitation or the provided "add as contacts" function; users can also tag/comment on the social contents like photos and videos; a user at YouTube can respond to another user by uploading a video; and a user can also become a fan of another user by subscription to the user's contributions of social contents. A multi-dimensional network among these users can be constructed based on all forms of activities, in which each dimension represents one facet of diverse interaction (Tang et al., 2009). A multi-dimensional network is also called *a multiplex network*, *multi-relational network*, or *labeled graph* in various domains.

- *Multi-Mode Networks*. A multi-mode network involves heterogeneous actors. Each mode represents one type of entity. A 3-mode network can be constructed for the aforementioned YouTube example, as seen in Figure 4.2. In the network, videos, tags and users each represent a

Figure 4.1: Various Types of Interactions at Content Sharing Sites

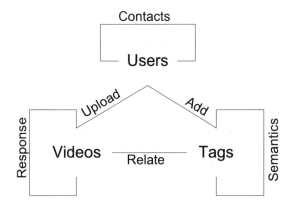

Figure 4.2: An Example of a Multi-Mode Network (based on (Tang et al., forthcoming))

mode. Note that both videos and tags are considered "actors" of the network as well, although users might be the major mode under consideration. Different interactions exist between the three types of entities: *users* can upload *videos*; users can add *tags* to a video. Videos and tags are naturally correlated to each other. Meanwhile, a friendship network exists among users, and a video clip can be uploaded to respond to another video. Tags can also connect to each other based on their semantic meanings. In other words, multiple types of entities exist in the same network, and entities relate to others (either the same type or different types) through different links. Besides social media, multi-mode networks are also observed in other domains such as information networks (Sun et al., 2009) from DBLP Bibliography[1] involving authors, papers and publication venues, or the Internet Movie Database[2] involving movies, actors, producers and directors.

Social media offers an easily-accessible platform for diverse online social activities, but it also introduces heterogeneity in networks. Thus, it calls for solutions to extract communities in

---

[1]http://www.informatik.uni-trier.de/~ley/db/
[2]http://www.imdb.com/

heterogeneous networks. However, it remains unanswered why one cannot reduce a heterogeneous network to several homogeneous ones (i.e., one mode or one dimension). The reason is that the interaction information in one mode or one dimension might be too noisy to detect meaningful communities. For instance, in the YouTube example in Figure 4.2, it seems acceptable if we only consider the user mode. In other words, just study the friendship network. On the one hand, some users might not have any online friends because they are too introvert to talk to other online users, or because they just join the network and are not ready for or not interested in connections. On the other hand, some users might abuse connections, since it is relatively easy to make connections in social media compared with in the physical world. For example, a user in Flickr can have thousands of friends (Tang and Liu, 2009b). This can hardly be true in the real world. It might be the case that two online users get connected but they never talk to each other. Thus, these online connections of one mode or one dimension alone can hardly paint a true picture of what is happening.

A single type of interaction provides limited information about the community membership of online users. Fortunately, social media provides more than just a single friendship network. A user might engage in other forms of activities besides connecting to friends. It is generally helpful to utilize information from other modes or dimensions for more effective community detection. Next, we will discuss how we may extend a community detection method to take into account interaction information at multiple dimensions or among multiple modes. For all the methods we discussed in this chapter, we assume the number of communities is given. The number of communities may be obtained via prior knowledge or by some exploratory study.

4.2   MULTI-DIMENSIONAL NETWORKS

Communications in social media are often multi-dimensional, leading to a multi-dimensional network. A p-dimensional network is represented as

$$\mathcal{A} = \{A^{(1)}, A^{(2)}, \cdots, A^{(p)}\},$$

with $A^{(i)} \in \{0, 1\}^{n \times n}$ represents the interaction among actors in the i-th dimension Through these interactions, a latent community structure emerges among actors. With multiple types of interactions, the shared latent community structure can be complicated. It is necessary to integrate information from multiple dimensions to find out the shared community structure across multiple network dimensions. In Figure 3.5, we show a unified process that summarizes existing community detection methods, involving four components: *network*, *utility matrix*, *soft community indicator*, and *partition*. Corresponding to each component, we can develop an integration strategy because soft[3] community indicators can be considered as structural features extracted from a network. Therefore, we have *network integration*, *utility integration*, *feature integration*, and *partition integration* as shown in Figure 4.3. We now delineate each integration strategy in detail. We use spectral clustering with normalized graph Laplacian (defined in Eq. (3.14)) as an example to go through the four strategies.

---

[3] *Soft* means we may have real values rather than boolean values to represent the degree of membership.

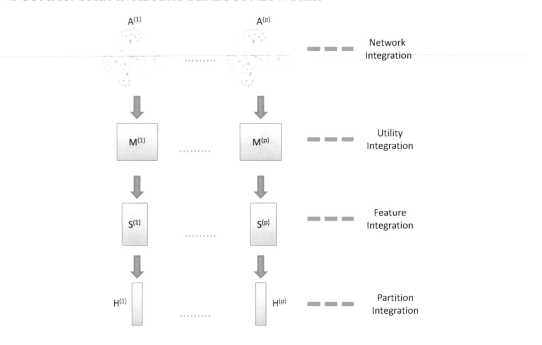

Figure 4.3: Multi-Dimensional Integration Strategies

Other variants of community detection methods (such as block models, latent space approach, and modularity maximization) in multi-dimensional networks (as suggested by the unified view) are also applicable (Tang, 2010).

4.2.1   NETWORK INTEGRATION

A simple strategy to handle a multi-dimensional network is to treat it as a single-dimensional network. The intuition here is that one's different interactions strengthen her connections in a mathematical sense of union. The average interaction among actors is:

$$\bar{A} = \frac{1}{p} \sum_{i=1}^{p} A^{(i)}. \tag{4.1}$$

With $\bar{A}$, this boils down to classical community detection in a single-dimensional network. Based on the average network, we can follow the community detection process as stated in the unified view. The objective function of spectral clustering with multi-dimensional networks should be changed accordingly:

$$\min_{S} Tr(S^T \bar{L} S) \ s.t. \ S^T S = I \ where \ \bar{L} = \bar{D}^{-1/2} \bar{A} \bar{D}^{-1/2}.$$

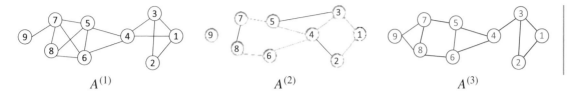

Figure 4.4: A 3-Dimensional Network

For the 3-dimensional network in Figure 4.4, we can compute the corresponding averaged adjacency matrix as follows:

$$\bar{A} = \begin{bmatrix} 0 & 1 & 1 & 2/3 & 0 & 0 & 0 & 0 & 0 \\ 1 & 0 & 2/3 & 1/3 & 0 & 0 & 0 & 0 & 0 \\ 1 & 2/3 & 0 & 1 & 1/3 & 0 & 0 & 0 & 0 \\ 2/3 & 1/3 & 1 & 0 & 1 & 1 & 0 & 0 & 0 \\ 0 & 0 & 1/3 & 1 & 0 & 2/3 & 1 & 1/3 & 0 \\ 0 & 0 & 0 & 1 & 2/3 & 0 & 1/3 & 1 & 0 \\ 0 & 0 & 0 & 0 & 1 & 1/3 & 0 & 1 & 2/3 \\ 0 & 0 & 0 & 0 & 1/3 & 1 & 1 & 0 & 1/3 \\ 0 & 0 & 0 & 0 & 0 & 0 & 2/3 & 1/3 & 0 \end{bmatrix}.$$

For example, between nodes 1 and 4, the total sum of interaction is 2. Hence, the average interaction should be 2/3. Based on the average similarity $\bar{A}$, we can compute the corresponding normalized graph Laplacian $\bar{L}$ and its top eigenvectors as follows:

$$\bar{L} = \begin{bmatrix} 1.00 & -0.43 & -0.35 & -0.20 & 0 & 0 & 0 & 0 & 0 \\ -0.43 & 1.00 & -0.27 & -0.12 & 0 & 0 & 0 & 0 & 0 \\ -0.35 & -0.27 & 1.00 & -0.29 & -0.11 & 0 & 0 & 0 & 0 \\ -0.20 & -0.12 & -0.29 & 1.00 & -0.27 & -0.29 & 0 & 0 & 0 \\ 0 & 0 & -0.11 & -0.27 & 1.00 & -0.21 & -0.32 & -0.11 & 0 \\ 0 & 0 & 0 & -0.29 & -0.21 & 1.00 & -0.11 & -0.35 & 0 \\ 0 & 0 & 0 & 0 & -0.32 & -0.11 & 1.00 & -0.35 & -0.38 \\ 0 & 0 & 0 & 0 & -0.11 & -0.35 & -0.35 & 1.00 & -0.20 \\ 0 & 0 & 0 & 0 & 0 & 0 & -0.38 & -0.20 & 1.00 \end{bmatrix}, \quad (4.2)$$

$$S = \begin{bmatrix} -0.33 & -0.44 \\ -0.28 & -0.40 \\ -0.35 & -0.38 \\ -0.40 & -0.18 \\ -0.37 & 0.16 \\ -0.35 & 0.21 \\ -0.35 & 0.41 \\ -0.33 & 0.38 \\ -0.20 & 0.30 \end{bmatrix}. \tag{4.3}$$

If k-means is applied to the structure feature S, we can partition the nodes into two groups $\{1, 2, 3, 4\}$ and $\{5, 6, 7, 8, 9\}$.

4.2.2   UTILITY INTEGRATION

Another variant for integration is to combine utility matrices instead of networks. We can obtain an average utility matrix as follows:

$$\bar{M} = \frac{1}{p} \sum_{i=1}^{p} M^{(i)},$$

where $M^{(i)}$ denotes the utility matrix constructed in the i-th dimension. The soft community indicators can be computed via the top eigenvectors of the utility matrix. This is equivalent to *optimizing the objective function over all types of interactions simultaneously*. As for spectral clustering, the average utility matrix in this case would be

$$\bar{M} = \frac{1}{p} \sum_{i=1}^{p} \widetilde{L}^{(i)}, \tag{4.4}$$

where $\widetilde{L}^{(i)}$ denote the normalized graph Laplacian of interaction at dimension i. Finding out the top eigenvectors of the average utility matrix is equivalent to minimizing the average normalized cut as follows:

$$\min_{S} \frac{1}{p} \sum_{i=1}^{p} Tr(S^T \widetilde{L}^{(i)} S) = \min_{S} Tr(S^T \bar{M} S). \tag{4.5}$$

Take the 3-dimensional network in Figure 4.4 as an example. The graph Laplacian for each type of interaction is listed below:

$$
\widetilde{L}^{(1)} = \begin{bmatrix}
1.00 & -0.41 & -0.33 & -0.29 & 0 & 0 & 0 & 0 & 0 \\
-0.41 & 1.00 & -0.41 & 0 & 0 & 0 & 0 & 0 & 0 \\
-0.33 & -0.41 & 1.00 & -0.29 & 0 & 0 & 0 & 0 & 0 \\
-0.29 & 0 & -0.29 & 1.00 & -0.25 & -0.25 & 0 & 0 & 0 \\
0 & 0 & 0 & -0.25 & 1.00 & -0.25 & -0.25 & -0.29 & 0 \\
0 & 0 & 0 & -0.25 & -0.25 & 1.00 & -0.25 & -0.29 & 0 \\
0 & 0 & 0 & 0 & -0.25 & -0.25 & 1.00 & -0.29 & -0.50 \\
0 & 0 & 0 & 0 & -0.29 & -0.29 & -0.29 & 1.00 & 0 \\
0 & 0 & 0 & 0 & 0 & 0 & -0.50 & 0 & 1.00
\end{bmatrix},
$$

$$
\widetilde{L}^{(2)} = \begin{bmatrix}
1.00 & -0.41 & -0.33 & -0.26 & 0 & 0 & 0 & 0 & 0 \\
-0.41 & 1.00 & 0 & -0.32 & 0 & 0 & 0 & 0 & 0 \\
-0.33 & 0 & 1.00 & -0.26 & -0.33 & 0 & 0 & 0 & 0 \\
-0.26 & -0.32 & -0.26 & 1.00 & -0.26 & -0.32 & 0 & 0 & 0 \\
0 & 0 & -0.33 & -0.26 & 1.00 & 0 & -0.41 & 0 & 0 \\
0 & 0 & 0 & -0.32 & 0 & 1.00 & 0 & -0.50 & 0 \\
0 & 0 & 0 & 0 & -0.41 & 0 & 1.00 & -0.50 & 0 \\
0 & 0 & 0 & 0 & 0 & -0.50 & -0.50 & 1.00 & 0 \\
0 & 0 & 0 & 0 & 0 & 0 & 0 & 0 & 0
\end{bmatrix},
$$

$$
\widetilde{L}^{(3)} = \begin{bmatrix}
1.00 & -0.50 & -0.41 & 0 & 0 & 0 & 0 & 0 & 0 \\
-0.50 & 1.00 & -0.41 & 0 & 0 & 0 & 0 & 0 & 0 \\
-0.41 & -0.41 & 1.00 & -0.33 & 0 & 0 & 0 & 0 & 0 \\
0 & 0 & -0.33 & 1.00 & -0.33 & -0.33 & 0 & 0 & 0 \\
0 & 0 & 0 & -0.33 & 1.00 & -0.33 & -0.33 & 0 & 0 \\
0 & 0 & 0 & -0.33 & -0.33 & 1.00 & 0 & -0.33 & 0 \\
0 & 0 & 0 & 0 & -0.33 & 0 & 1.00 & -0.33 & -0.41 \\
0 & 0 & 0 & 0 & 0 & -0.33 & -0.33 & 1.00 & -0.41 \\
0 & 0 & 0 & 0 & 0 & 0 & -0.41 & -0.41 & 1.00
\end{bmatrix}.
$$

It follows that $\bar{M} = \left(\widetilde{L}^{(1)} + \widetilde{L}^{(2)} + \widetilde{L}^{(3)}\right)/3$. Thus,

$$\bar{M} = \begin{bmatrix} 1.00 & -0.44 & -0.36 & -0.18 & 0 & 0 & 0 & 0 & 0 \\ -0.44 & 1.00 & -0.27 & -0.11 & 0 & 0 & 0 & 0 & 0 \\ -0.36 & -0.27 & 1.00 & -0.29 & -0.11 & 0 & 0 & 0 & 0 \\ -0.18 & -0.11 & -0.29 & 1.00 & -0.28 & -0.30 & 0 & 0 & 0 \\ 0 & 0 & -0.11 & -0.28 & 1.00 & -0.19 & -0.33 & -0.10 & 0 \\ 0 & 0 & 0 & -0.30 & -0.19 & 1.00 & -0.08 & -0.37 & 0 \\ 0 & 0 & 0 & 0 & -0.33 & -0.08 & 1.00 & -0.37 & -0.30 \\ 0 & 0 & 0 & 0 & -0.10 & -0.37 & -0.37 & 1.00 & -0.14 \\ 0 & 0 & 0 & 0 & 0 & 0 & -0.30 & -0.14 & 0.67 \end{bmatrix}. \quad (4.6)$$

Clearly, this average Laplacian is different from the Laplacian constructed on average network interactions (Eq. (4.2)). The smallest two eigenvectors with respect to $\bar{M}$ are

$$S = \begin{bmatrix} 0.33 & 0.43 \\ 0.29 & 0.39 \\ 0.36 & 0.37 \\ 0.41 & 0.17 \\ 0.37 & -0.15 \\ 0.34 & -0.19 \\ 0.34 & -0.40 \\ 0.32 & -0.37 \\ 0.22 & -0.38 \end{bmatrix}.$$

In this example, the soft community indicator of the utility interaction is similar to that based on network utility (Eq. (4.3)), except for signs. A partition of two communities ({1, 2, 3, 4} and {5, 6, 7, 8, 9}) is already encoded in the second column of S by the sign.

4.2.3 FEATURE INTEGRATION

Soft community indicators extracted from each dimension of the network are structural features associated with nodes. When we try to integrate them, the integration is performed at the feature level, thus *feature integration*. One might conjecture that we can perform similar operations as we did for network interactions and utility matrices, i.e., taking the average of structural features as follows:

$$\bar{S} = \frac{1}{p} \sum_{i=1}^{p} S^{(i)}. \quad (4.7)$$

Unfortunately, this straightforward average does not apply to structural features. The solution S that optimizes the utility function is not unique. Dissimilar structural features may not necessarily suggest differences of the corresponding latent community structures. In the simplest case, $S' = -S$ is also a valid solution. Averaging these structural features does not result in sensible features.

To give a palpable understanding, we can take a look at the 3-dimensional network in Figure 4.4. For presentation convenience, we extract 3 soft community indicators for each type of interaction. The corresponding structural features are list below:

$$
S^{(1)} = \begin{bmatrix}
0.33 & -0.44 & 0.09 \\
0.27 & -0.43 & 0.22 \\
0.33 & -0.44 & 0.09 \\
0.38 & -0.16 & -0.32 \\
0.38 & 0.24 & -0.30 \\
0.38 & 0.24 & -0.30 \\
0.38 & 0.38 & 0.42 \\
0.33 & 0.30 & -0.16 \\
0.19 & 0.23 & 0.67
\end{bmatrix}, \quad
S^{(2)} = \begin{bmatrix}
-0.37 & 0 & 0.39 \\
-0.30 & 0 & 0.33 \\
-0.37 & 0 & 0.23 \\
-0.48 & 0 & 0.21 \\
-0.37 & 0 & -0.08 \\
-0.30 & 0 & -0.31 \\
-0.30 & 0 & -0.46 \\
-0.30 & 0 & -0.56 \\
0 & 1.00 & 0
\end{bmatrix},
$$

$$
S^{(3)} = \begin{bmatrix}
0.29 & 0.47 & 0.21 \\
0.29 & 0.47 & 0.21 \\
0.35 & 0.44 & 0.01 \\
0.35 & 0.04 & -0.50 \\
0.35 & -0.17 & -0.39 \\
0.35 & -0.17 & -0.39 \\
0.35 & -0.33 & 0.28 \\
0.35 & -0.33 & 0.28 \\
0.29 & -0.30 & 0.45
\end{bmatrix}.
$$

If we simply take the average as in Eq. (4.7), we obtain $\bar{S}$ as follows:

$$
\bar{S} = \begin{bmatrix}
0.08 & 0.01 & 0.23 \\
0.08 & 0.01 & 0.26 \\
0.10 & 0.00 & 0.11 \\
0.08 & -0.04 & -0.20 \\
0.12 & 0.02 & -0.26 \\
0.14 & 0.02 & -0.34 \\
0.14 & 0.02 & 0.08 \\
0.13 & -0.01 & -0.15 \\
0.16 & 0.31 & 0.37
\end{bmatrix}.
$$

Note that none of the three columns contains strong community information. If we apply k-means to $\bar{S}$ directly, we obtain the following partition of nodes: $\{1, 2, 3, 7, 9\}$ and $\{4, 5, 6, 8\}$, which does not make much sense.

Let $S_\ell^{(i)}$ denote the ℓ-th column of $S^{(i)}$. We can pay special attention of the case when $\ell = 2$ for all $S^{(i)}$. Though both $S_2^{(1)}$ and $S_2^{(3)}$ imply a partition of separating nodes $\{1, 2, 3, 4\}$ and

---

Algorithm: Structural Feature Integration

    Input: $Net = \{A^{(1)}, A^{(2)}, \cdots, A^{(p)}\}$,
           number of communities k,
           number of structural features to extract ℓ;
    Output: community partition H.

---

1. Compute top ℓ eigenvectors of the utility matrix
2. Compute slim SVD of $X = [S^{(1)}, S^{(2)}, \cdots S^{(p)}] = UDV^T$;
3. Obtain averaged structural features $\bar{S}$ as the first k columns of U;
4. Calculate the community partition with k-means on $\bar{S}$.

Figure 4.5: Structural Feature Integration Algorithm for Multi-Dimensional Networks

$\{5, 6, 7, 8, 9\}$ (as indicated by the sign in one column), their summation cancels each other, leading to a sum close to 0. On the other hand, only node 9 is assigned with a non-zero value in $S_2^{(2)}$, indicating the community as a singleton: node 9 is isolated in $A^{(2)}$. When taking the average, the second column of $\bar{S}$ does not contain much meaningful information.

Alternatively, we expect structural features of different dimensions to be highly correlated *after a certain transformation* (Long et al., 2008). For instance, $S_2^{(1)}$ and $S_2^{(3)}$ are highly correlated; $S_2^{(1)}$ is also correlated with $S_3^{(2)}$. The multi-dimensional integration can be conducted after we apply a transformation to structural features to map them into the same coordinates (Tang et al., 2009). In order to find out the transformation associated with each type of interaction (say, $\mathbf{w}^{(i)}$), one can maximize the pairwise correlations between structural features extracted from different types of interactions. Once such a transformation is found, it can be shown that the averaged structural features after transformation $\bar{S}$, i.e., $\frac{1}{p} \sum_{i=1}^{p} S^{(i)} \mathbf{w}^{(i)}$, is proportional to the left singular vectors of the following data:

$$X = \left[S^{(1)}, S^{(2)}, \cdots, S^{(p)} \right].$$

Hence, we have a structural feature integration algorithm as in Figure 4.5. In summary, we first extract structural features from each dimension of the network via a community detection method; then SVD is applied to the concatenated data $X = [S^{(1)}, S^{(2)}, \cdots S^{(p)}]$. The average feature $\bar{S}$ across all dimensions of a network is captured by the left singular vectors of X. Then we can perform k-means on this averaged structural feature $\bar{S}$ to find a discrete community partition.

Let us re-examine the 3-dimensional network example. We can concatenate all structural features to form X.

$$X = \begin{bmatrix}
0.33 & -0.44 & 0.09 & -0.37 & 0 & 0.39 & 0.29 & 0.47 & 0.21 \\
0.27 & -0.43 & 0.22 & -0.30 & 0 & 0.33 & 0.29 & 0.47 & 0.21 \\
0.33 & -0.44 & 0.09 & -0.37 & 0 & 0.23 & 0.35 & 0.44 & 0.01 \\
0.38 & -0.16 & -0.32 & -0.48 & 0 & 0.21 & 0.35 & 0.04 & -0.50 \\
0.38 & 0.24 & -0.30 & -0.37 & 0 & -0.08 & 0.35 & -0.17 & -0.39 \\
0.38 & 0.24 & -0.30 & -0.30 & 0 & -0.31 & 0.35 & -0.17 & -0.39 \\
0.38 & 0.38 & 0.42 & -0.30 & 0 & -0.46 & 0.35 & -0.33 & 0.28 \\
0.33 & 0.30 & -0.16 & -0.30 & 0 & -0.56 & 0.35 & -0.33 & 0.28 \\
0.19 & 0.23 & 0.67 & 0 & 1.00 & 0 & 0.29 & -0.30 & 0.45
\end{bmatrix}.$$

The top 2 left singular vectors of X are listed below:

$$\bar{S} = \begin{bmatrix}
-0.30 & 0.42 \\
-0.26 & 0.38 \\
-0.33 & 0.38 \\
-0.39 & 0.24 \\
-0.37 & -0.07 \\
-0.36 & -0.14 \\
-0.36 & -0.40 \\
-0.35 & -0.36 \\
-0.23 & -0.40
\end{bmatrix}.$$

Notice that this averaged partition information contains the partition information in the second column. A partition of $\{1, 2, 3, 4\}$ and $\{5, 6, 7, 8, 9\}$ can be obtained after k-means clustering on $\bar{S}$.

4.2.4   PARTITION INTEGRATION

Partition integration takes effect after the community partition of each network dimension is ready. This problem has been studied as the *cluster ensemble* problem (Strehl and Ghosh, 2003), which combines multiple clustering results of the same data from a variety of sources into a single consensus clustering. Strehl and Ghosh (2003) propose three effective and comparable approaches: cluster-based similarity partitioning algorithm (CPSA), HyperGraph Partition Algorithm and Meta-Clustering Algorithm. For brevity, we only present the basic idea of CPSA here. CPSA constructs a similarity matrix from each clustering. Two objects' similarity is 1 if they belong to the same group, 0 if they belong to different groups. Let $H^{(i)} \in \{0, 1\}^{n \times k}$ denote the community indicator matrix of clustering based on interactions at dimension i. The similarity between nodes can be computed as

$$\frac{1}{p} \sum_{i=1}^{p} H^{(i)} (H^{(i)})^T = \frac{1}{p} \sum_{i=1}^{p} Y Y^T \text{ where } Y = \left[H^{(1)}, H^{(2)}, \cdots, H^{(p)} \right].$$

This basically amounts to the probability that two nodes are assigned into the same community. Based on this similarity matrix between nodes, we can apply similarity-based community detection methods we introduced before to find out clusters.

Let us look at the 3-dimensional example again. Suppose we partition each network dimension into two communities, we obtain the following partition:

$$
H^{(1)} = \begin{bmatrix} 1 & 0 \\ 1 & 0 \\ 1 & 0 \\ 1 & 0 \\ 0 & 1 \\ 0 & 1 \\ 0 & 1 \\ 0 & 1 \\ 0 & 1 \end{bmatrix}, \quad
H^{(2)} = \begin{bmatrix} 1 & 0 \\ 1 & 0 \\ 1 & 0 \\ 1 & 0 \\ 1 & 0 \\ 1 & 0 \\ 1 & 0 \\ 1 & 0 \\ 0 & 1 \end{bmatrix}, \quad
H^{(3)} = \begin{bmatrix} 1 & 0 \\ 1 & 0 \\ 1 & 0 \\ 0 & 1 \\ 0 & 1 \\ 0 & 1 \\ 0 & 1 \\ 0 & 1 \\ 0 & 1 \end{bmatrix}.
$$

It follows that

$$
\frac{1}{p}\sum_{i=1}^{p} H^{(i)}(H^{(i)})^T = \begin{bmatrix}
1.00 & 1.00 & 1.00 & 0.67 & 0.33 & 0.33 & 0.33 & 0.33 & 0 \\
1.00 & 1.00 & 1.00 & 0.67 & 0.33 & 0.33 & 0.33 & 0.33 & 0 \\
1.00 & 1.00 & 1.00 & 0.67 & 0.33 & 0.33 & 0.33 & 0.33 & 0 \\
0.67 & 0.67 & 0.67 & 1.00 & 0.67 & 0.67 & 0.67 & 0.67 & 0.33 \\
0.33 & 0.33 & 0.33 & 0.67 & 1.00 & 1.00 & 1.00 & 1.00 & 0.67 \\
0.33 & 0.33 & 0.33 & 0.67 & 1.00 & 1.00 & 1.00 & 1.00 & 0.67 \\
0.33 & 0.33 & 0.33 & 0.67 & 1.00 & 1.00 & 1.00 & 1.00 & 0.67 \\
0.33 & 0.33 & 0.33 & 0.67 & 1.00 & 1.00 & 1.00 & 1.00 & 0.67 \\
0 & 0 & 0 & 0.33 & 0.6667 & 0.67 & 0.67 & 0.67 & 1.00
\end{bmatrix}. \quad (4.8)
$$

This new adjacency matrix can be considered as a network with each entry being the weight of connectivity. The community detection methods discussed in the previous chapter can be applied to this weighted network to find out partitions. For example, we can apply spectral clustering to this new network and then obtain a partition of two communities $\{1, 2, 3, 4\}$ and $\{5, 6, 7, 8, 9\}$.

A disadvantage of this CPSA is that the computed similarity matrix can be dense, which might not be applicable to large networks. Instead, we can treat Y as the feature representation of actors and apply a similar procedure as in feature integration. That is, we compute the top left singular vectors of Y and cluster nodes via k-means on the extracted singular vectors. As for the

	Network Integration	Utility Integration	Feature Integration	Partition Integration
Table 4.1: Comparing Different Multi-Dimensional Integration Strategies				
Tuning weights for different interactions	✓	✓	✓	✓
Sensitivity to noise	yes	ok	robust	yes
Clustering quality	bad	good	good	ok
Computational cost	low	low	high	expensive

3-dimensional network, we have Y as follows:

$$Y = \begin{bmatrix} 1 & 0 & 1 & 0 & 1 & 0 \\ 1 & 0 & 1 & 0 & 1 & 0 \\ 1 & 0 & 1 & 0 & 1 & 0 \\ 1 & 0 & 1 & 0 & 0 & 1 \\ 0 & 1 & 1 & 0 & 0 & 1 \\ 0 & 1 & 1 & 0 & 0 & 1 \\ 0 & 1 & 1 & 0 & 0 & 1 \\ 0 & 1 & 1 & 0 & 0 & 1 \\ 0 & 1 & 0 & 1 & 0 & 1 \end{bmatrix}.$$

Its top 2 left singular vectors are

$$\bar{H} = \begin{bmatrix} -0.27 & -0.47 \\ -0.27 & -0.47 \\ -0.27 & -0.47 \\ -0.35 & -0.14 \\ -0.39 & 0.22 \\ -0.39 & 0.22 \\ -0.39 & 0.22 \\ -0.39 & 0.22 \\ -0.24 & 0.35 \end{bmatrix}.$$

Clearly, the second column of H_2 encodes the partition information, which can be easily identified by k-means. In this way, we avoid the construction of a huge dense matrix as in Eq. (4.8).

All the four integration strategies can be extended by adding regularization or assigning varying weights to different types of interactions. It has been shown that utility integration and feature integration tend to outperform simple network integration or partition integration (Tang et al., 2009). Among the four strategies, utility integration typically outputs reasonably good performance. Structural feature integration involves computing soft community indicators from each type of interaction. This step can be considered as a denoise process, thus the subsequent integration is

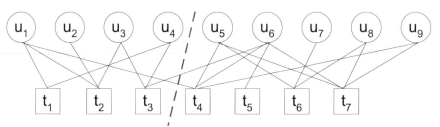

Figure 4.6: A two-mode network represented as a bipartite graph. The dashed line represents one normalized cut found by co-clustering.

able to extract a more accurate community structure. In terms of computational cost, they can be sorted from low to high as follows: network integration, utility integration, feature integration, and partition integration. The pros and cons of different multi-dimensional integration strategies are summarized in Table 4.1. Given their advantage in clustering performance, utility integration or feature integration should be preferred.

4.3   MULTI-MODE NETWORKS

So far, we mainly focused on networks with only one type of entity. In reality, entities of distinctive types can interact also with each other. For instance, for a bibliography network, we have authors, publication venues, papers, terms in titles fabricated into one network. In social media, a network can be composed of multiple types of entities. The network is called a multi-mode network. We start with a two-mode network - the simplest form of multi-mode networks, and discuss how co-clustering works. Then, we generalize the algorithm to multi-mode networks.

4.3.1   CO-CLUSTERING ON TWO-MODE NETWORKS

In some content-sharing sites, users upload tags to shared content such as videos, photos and bookmarks, which naturally forms a user-tag network. In this network, we have two types of entities: users and tags, and the connection is often between a user and a tag. Users are indirectly connected via tags. It is a *two-mode network*, with one mode representing one type of entity. In traditional social science, a two-mode network is also named as an *affiliation network* (Wasserman and Faust, 1994).

Akin to typical one-mode networks, there are also two ways to represent a two-mode network: a graph or an adjacency matrix. Figure 4.6 shows an example of a two-mode network between 9 users and 7 tags. Clearly, the graph is *bipartite* because all the edges are between users and tags and none between users or between tags. The corresponding adjacency matrix of the network is shown in Table 4.2. Since each mode might contain a different number of entities (9 for the user mode, and 7 for the tag mode), the corresponding adjacency matrix is not a square matrix.

In a two-mode network, communities in one mode become correlated with communities in the other mode. Communities in one mode are formed due to their members' shared connections

	t_1	t_2	t_3	t_4	t_5	t_6	t_7
				Table 4.2: Adjacency Matrix			
u_1	1	1	0	1	0	0	0
u_2	0	1	0	0	0	0	0
u_3	0	1	1	0	0	0	0
u_4	1	0	1	0	0	0	0
u_5	0	0	0	1	0	1	1
u_6	0	0	1	1	1	0	1
u_7	0	0	0	0	0	1	0
u_8	0	0	0	0	0	1	1
u_9	0	0	0	1	0	0	1

to those in the other mode. For example, in a user-tag network, users sharing similar interests are likely to use similar tags, and semantically similar tags tend to be used by users of similar interests. Thus, *co-clustering* (Dhillon, 2001) is proposed to find communities in two modes simultaneously. Co-clustering is also known as *biclustering* (Madeira and Oliveira, 2004). For a user-tag network, co-clustering not only outputs communities about users, but also communities about tags.

There are many approaches to perform co-clustering (Madeira and Oliveira, 2004). Here, we illustrate a spectral clustering approach extended to two-mode networks. We can again minimize the cut in the graph of a two-mode network, which might lead to a highly imbalanced cut. For instance, in the network in Figure 4.6, ignoring the entity type of modes, the minimum cut is 1, between $\{u_2\}$ and the remaining network, or t_5 and the rest, or u_7 and the rest. In all the cases, we obtain a community composed of only one node (u_2, t_5, u_7, respectively for each case). Alternatively, the dashed line in Figure 4.6 seems a more reasonable cut. We can optimize the ratio cut or the normalized cut to take into account the size of a community as we discussed in Section 3.3.4.

The spectral co-clustering algorithm aims to minimize the normalized cut in a bipartite graph. It proceeds as follows:

- Given a two-mode network A, normalize the adjacency matrix as

$$\widetilde{A} = D_u^{-1/2} A D_t^{-1/2}, \tag{4.9}$$

where

$$D_u = diag(d_{u_1}, d_{u_2}, \cdots, d_{u_m}), \quad D_t = diag(d_{t_1}, d_{t_2}, \cdots, d_{t_n})$$

with d_{u_i} and d_{t_j} representing the degree of u_i and t_j, respectively.

- Compute the SVD of $\widetilde{A}$, say $\widetilde{A} = U \Sigma V^T$. Let $S^{(u)} = U_{1:\ell}$, $S^{(t)} = V_{1:\ell}$, where ℓ is typically set to k if we aim to find k communities. Then, $S^{(u)}$ is the soft community indicator of the user mode, and $S^{(t)}$ the soft community indicator of the tag mode.

- After we obtain $S^{(u)}$ and $S^{(t)}$, k-means is run to obtain communities. We obtain a joint soft community indicator of both the user mode and the tag mode as Z:

$$Z = \begin{bmatrix} D_u^{-1/2} S^{(u)} \\ D_t^{-1/2} S^{(t)} \end{bmatrix}.$$

Communities can be obtained by applying k-means to this joint soft community indicator Z.

The final algorithm of co-clustering is similar to spectral clustering. Rather than computing the smallest of eigenvectors of a graph Lapacian, we compute the singular value decomposition (SVD) of the normalized adjacency matrix of a given two-mode network.

In this co-clustering algorithm, we run k-means algorithm on the reduced representation Z of users and tags simultaneously. For example, for the network in Figure 4.6,

$$\begin{aligned} D_u &= diag(3, 1, 2, 2, 3, 4, 1, 2, 2), \\ D_t &= diag(2, 3, 3, 4, 1, 3, 4). \end{aligned}$$

Thus, we obtained a normalized adjacency matrix as follows:

$$\widetilde{A} = D_u^{-1/2} A D_t^{-1/2} = \begin{bmatrix} 0.41 & 0.33 & 0 & 0.29 & 0 & 0 & 0 \\ 0 & 0.58 & 0 & 0 & 0 & 0 & 0 \\ 0 & 0.41 & 0.41 & 0 & 0 & 0 & 0 \\ 0.50 & 0 & 0.41 & 0 & 0 & 0 & 0 \\ 0 & 0 & 0 & 0.29 & 0 & 0.33 & 0.29 \\ 0 & 0 & 0.29 & 0.25 & 0.50 & 0 & 0.25 \\ 0 & 0 & 0 & 0 & 0 & 0.58 & 0 \\ 0 & 0 & 0 & 0 & 0 & 0.41 & 0.36 \\ 0 & 0 & 0 & 0.36 & 0 & 0 & 0.36 \end{bmatrix}.$$

If we partition users and tags in the network into two communities, the top two left and right singular vectors and the corresponding Z are:

$$
S^{(u)} = \begin{bmatrix}
-0.39 & 0.33 \\
-0.22 & 0.35 \\
-0.32 & 0.40 \\
-0.32 & 0.35 \\
-0.39 & -0.37 \\
-0.45 & -0.04 \\
-0.22 & -0.38 \\
-0.32 & -0.42 \\
-0.32 & -0.19
\end{bmatrix}, \quad
S^{(t)} = \begin{bmatrix}
-0.32 & 0.35 \\
-0.39 & 0.35 \\
-0.39 & 0.33 \\
-0.45 & -0.10 \\
-0.22 & -0.02 \\
-0.39 & -0.58 \\
-0.45 & -0.38
\end{bmatrix}, \quad
Z = \begin{bmatrix}
u_1 & -0.22 & 0.19 \\
u_2 & -0.22 & 0.35 \\
u_3 & -0.22 & 0.28 \\
u_4 & -0.22 & 0.25 \\
u_5 & -0.22 & -0.22 \\
u_6 & -0.22 & -0.02 \\
u_7 & -0.22 & -0.38 \\
u_8 & -0.22 & -0.29 \\
u_9 & -0.22 & -0.14 \\
t_1 & -0.22 & 0.25 \\
t_2 & -0.22 & 0.32 \\
t_3 & -0.22 & 0.19 \\
t_4 & -0.22 & -0.05 \\
t_5 & -0.22 & -0.02 \\
t_6 & -0.22 & -0.33 \\
t_7 & -0.22 & -0.19
\end{bmatrix}.
$$

The first vector of Z assigns all users and tags to the same community, which does not provide much useful information, thus is typically discarded. If we run k-means with $k = 2$ on the remaining column of Z, we obtain two communities $\{u_1, u_2, u_3, u_4, t_1, t_2, t_3\}$ and $\{u_5, u_6, u_7, u_8, u_9, t_4, t_5, t_6, t_7\}$, which is exactly the cut as indicated by the dashed line in Figure 4.6.

4.3.2   GENERALIZATION TO MULTI-MODE NETWORKS

Multi-mode networks involve multiple types of entities, or multiple modes. In a multi-mode network, each pair of modes might interact with each other. We may also have within-mode interactions. How could we extend the previous co-clustering algorithm to handle multi-mode networks?

Clearly, one key step in this spectral co-clustering algorithm is the SVD of $\widetilde{A}$. This step can also be interpreted as a block model approximation (Section 3.3.3) to a normalized two-mode network, because the singular value decomposition is well known to be an optimal solution to the following formula (Golub and Van Loan, 1996):

$$
\min_{S^{(1)}, S^{(2)}, \Sigma} \|\widetilde{A} - S^{(1)} \Sigma S^{(2)}\|_F^2, \quad s.t. \ (S^{(1)})^T S^{(1)} = I_k, \ (S^{(2)})^T S^{(2)} = I_k, \tag{4.10}
$$

where $S^{(1)}$ and $S^{(2)}$ act as the soft community indicators in corresponding modes. Based on the same criterion, we can generalize the formula to handle a multi-mode network. For simplicity, let us consider a special case where the multi-mode network is a star structure as in Figure 4.7. As we will

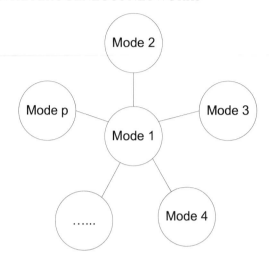

Figure 4.7: A Star Structure Concerning Mode 1

show later, a more general multi-mode network can be decomposed into multiple star-structures with respect to each mode.

Suppose there are p modes, with mode 1 interacting with the other $p - 1$ modes. Let the interaction between mode 1 and mode q as $A^{(q)}$. Thus, we can minimize the overall block model approximation error:

$$\min \quad \sum_{q=2}^{p} \|\widetilde{A}^{(q)} - S^{(1)}\Sigma^{(q)}(S^{(q)})^T\|_F^2, \tag{4.11}$$

$$s.t. \quad (S^{(1)})^T S^{(1)} = I_k, \quad (S^{(q)})^T S^{(q)} = I_k, \quad q = 2, \cdots, p. \tag{4.12}$$

In the formula, $\widetilde{A}^{(q)}$ denotes the interaction between mode 1 and mode q, $S^{(q)}$ the soft community indicator of mode q, and $\Sigma^{(q)}$ can be roughly considered as the inter-block density of $\widetilde{A}^{(q)}$. It can be shown that when $\{S^{(q)}|q = 2, \cdots, k\}$ are fixed, $S^{(1)}$ can be computed as the left singular vectors of the following matrix (Long et al., 2006; Tang et al., 2008):

$$X = \left[\widetilde{A}^{(2)}V^{(2)}, \widetilde{A}^{(3)}V^{(3)}, \cdots, \widetilde{A}^{(p)}V^{(p)} \right]. \tag{4.13}$$

Or equivalently, the eigenvectors of the matrix $M = XX^T$. The matrix X can be considered as feature representation extracted from related modes, and M is the similarity matrix of nodes in mode 1. Thus, if within-mode interaction (denoted as $A^{(1)}$) is also available, we can simply add the normalized within-mode interaction $\widetilde{A}^{(1)}$ to M, i.e., $M = XX^T + \widetilde{A}^{(1)}$.

As for a general multi-mode network, we can compute the soft community indicator of each mode one by one. If we look at one mode versus other modes, it is essentially a star structure.

Given the soft community indicators of its related modes, the soft community of one mode can be updated using SVD as we described in Eq. (4.13). In sum, in order to find communities at each mode, we first normalize all the interaction matrix in a multi-mode network following Eq. (4.9). Then, we iteratively update communities at each mode following iterative SVD. After we obtain soft community indicators for each mode, k-means can be applied to find a disjoint partition.

The algorithm we presented here for community detection in multi-mode networks is somehow simplified. Many extensions can be added such as assigning different weights to between-mode interactions and a varying number of communities in different modes. Interesting readers can refer to (Tang et al., forthcoming) for further readings.

CHAPTER 5

Social Media Mining

In previous chapters, we have discussed various approaches to community detection in social media. But after we obtain communities, what can we do with the extracted communities? In this chapter, we discuss two applications of social media mining. One is to study community evolution patterns in social media, and the other is to leverage social media networks to predict user behaviors.

5.1 EVOLUTION PATTERNS IN SOCIAL MEDIA

In social media, networks are highly dynamic. Each day, new members might join a network and new connections are established. Some existing members might become dormant as well. This yields a highly dynamic network. Kumar et al. (2005) show that some statistics such as degrees, the size of maximum connected component, and the number of communities observe a significant change around 2001 in the blogosphere. Some social media sites also observe a tremendous growth in recent years, e.g., the number of active users in Facebook grows from 1 million in 2004 to 500 million in 2010 as shown in Figure 5.1. Such a tremendous growth of networks provides unprecedented opportunities to study network dynamics.

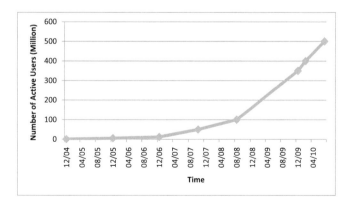

Figure 5.1: Growth of Active Users on Facebook

One fundamental question to understand network dynamics is how new connections are created in a dynamic network. How do network statistics change such as the size of the giant connected component, the diameter of a network, or the density of connections? The recent availability of many grand-size dynamic networks in social media allows researchers to investigate properties in evolving networks for new insights. Kossinets and Watts (2006) investigate email communications of students in a university over time, and they discover that the probability of a new connection between two nodes is disproportionate to their social distance. This phenomenon of decreasing probability of new connections between two nodes with increasing distance between them, is also confirmed later in (Leskovec et al., 2008). A large portion of new connections occur in a form of *triadic closures*. That is, two actors are likely to become friends when they share common friends. Kumar et al. (2006), by studying Flickr network comprising of 1 million nodes and 8 million directed edges spanning from Flickr's public launch in 2004 to 2006, and another social networking site Yahoo! 360 (now stopped), observe a segmentation of the networks into three regions: singletons who do not participate in the network, isolated communities which overwhelmingly display star structure, and a giant component anchored by a well-connected core region which persists even in the absence of stars. Leskovec et al. (2007a) study a wide range of graphs and find out that 1) most networks become more and more dense with new nodes arrival, and 2) the average distance between nodes shrinks over time.

With the growth of networks, communities can also expand, shrink, or dissolve. As argued by Kumar et al. (2005), communities in social media, in particular blogspace at that moment, are different from conventional web communities. "Within a community of interacting bloggers, a given topic may become the subject of intense debate for a period of time, then fade away. These bursts of activity are typified by heightened hyperlinking amongst the blogs involved—within a time interval." By studying the explicit group subscription of users in a popular blogging site LiveJournal[1], and the co-authorship network from DBLP[2], Backstrom et al. (2006) unravel the structural feature that is most important in determining whether a user joins a group. It is the number of friends that are already in the group affecting most in guiding an actor to join a group. Latent communities resulting from noisy network interactions are to evolve as well, which adds a new temporal dimension to the problem of community detection. We now discuss issues of finding community evolution in dynamic networks

5.1.1   A NAIVE APPROACH TO STUDYING COMMUNITY EVOLUTION

In order to discover community evolution, one simple approach is to take a series of snapshots of a network, and apply a community detection method (refer to Chapter 3) to each snapshot. Since clustering is applied independently in each snapshot, it is called an *independent clustering* approach. By comparing communities across time, different events concerning groups can be defined. These events include growth, contraction, merging, splitting, birth, and death as shown in Figure 5.2 (Palla et al., 2007).

A general process to analyze community evolution (Asur et al., 2007) is presented in Figure 5.3. Given a sequence of network snapshots, we can conduct community detection at each individual snapshot. Then, event detection or other behavioral analysis can be conducted in order to mine the evolution patterns of communities. As indicated by the general process, community

---

[1]http://www.livejournal.com/
[2]http://www.informatik.uni-trier.de/~ley/db/

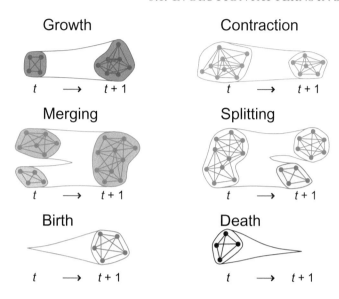

Figure 5.2: Events in Community Evolution (based on (Palla et al., 2007))

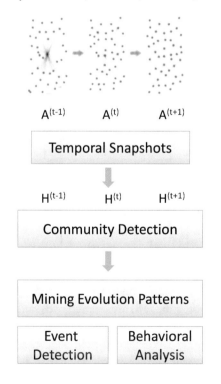

Figure 5.3: Mining Community Evolution in Dynamic Networks

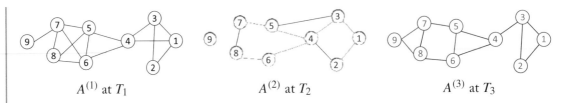

Figure 5.4: Three Consecutive Network Snapshots

detection can be applied to each snapshot to find out latent community structures. Such a scheme has been applied in (Hopcroft et al., 2004; Kumar et al., 2005; Palla et al., 2007; Asur et al., 2007).

Most community detection methods, however, suffer from local optima. A minute change of network connection, algorithm initialization, or the processing order of nodes might lead to a dramatically different community structure. For example, Hopcroft et al. (2003) show that the community structure based on hierarchical agglomerative clustering can change sharply with the removal of one single edge in a network. Some extracted communities are fairly random. With a community detection method that is unstable or has multiple local optimal solutions, it is difficult to conclude whether there is a community evolution between consecutive timestamps. It could simply be the randomness inherited in the chosen algorithm that causes the structural change.

For example, given three consecutive network snapshots in Figure 5.4, if we apply spectral clustering to each network snapshot to find two communities, we obtain the following community structures $H^{(1)}$, $H^{(2)}$ and $H^{(3)}$ at T_1, T_2 and T_3, respectively.

$$
H^{(1)} = \begin{bmatrix} 1 & 0 \\ 1 & 0 \\ 1 & 0 \\ 1 & 0 \\ 0 & 1 \\ 0 & 1 \\ 0 & 1 \\ 0 & 1 \\ 0 & 1 \end{bmatrix}, \quad
H^{(2)} = \begin{bmatrix} 1 & 0 \\ 1 & 0 \\ 1 & 0 \\ 1 & 0 \\ 1 & 0 \\ 1 & 0 \\ 1 & 0 \\ 1 & 0 \\ 0 & 1 \end{bmatrix}, \quad
H^{(3)} = \begin{bmatrix} 1 & 0 \\ 1 & 0 \\ 1 & 0 \\ 0 & 1 \\ 0 & 1 \\ 0 & 1 \\ 0 & 1 \\ 0 & 1 \\ 0 & 1 \end{bmatrix}.
$$

Based on the derived community partition, it seems that at time T_2, there is a sharp change. Nodes $\{5, 6, 7, 8\}$ are merged into the community of $\{1, 2, 3, 4\}$. While at time T_3, it seems the structure is changed again. Nodes $\{4, 5, 6, 7, 8\}$ are merged with node 9 and separated from nodes $\{1, 2, 3\}$. This shows a potential drawback of independent clustering: it can suggest spurious structural changes.

Hence, if clustering is applied to each snapshot independently, a *robust* method has to be used in order to derive meaningful evolution patterns. Palla et al. (2007), for instance, adopt a clique percolation method (Palla et al., 2005) to find communities in each snapshot. The clique percolation method outputs a unique community structure given a network and pre-specified parameters.

Hopcroft et al. (2004) resort to *natural communities*. They are defined as a set of nodes that are often grouped together across multiple runs of a clustering algorithm on a network with a small portion of connections randomly removed. These natural communities are more robust with respect to a random change in networks or clustering algorithms. In sum, a robust clustering method should be adopted. Otherwise, the discovered evolution pattern could be misleading.

5.1.2   COMMUNITY EVOLUTION IN SMOOTHLY EVOLVING NETWORKS

Another approach to find latent community structure at each snapshot is *evolutionary clustering* (Chakrabarti et al., 2006). Evolutionary clustering hinges on the assumption of *temporal smoothness*, i.e., the community structure does not change abruptly between consecutive timestamps. This temporal smoothness can be exploited to overcome the randomness introduced by network noise or community detection methods.

By contrast, evolutionary clustering aims to find a *smooth* sequence of communities given a series of network snapshots. The overall objective function of evolutionary clustering can be decomposed into *snapshot cost* (denoted as CS) and *temporal cost* (denoted as CT) (Chi et al., 2007):

$$Cost = \alpha \cdot CS + (1 - \alpha) \cdot CT. \tag{5.1}$$

The snapshot cost measures the quality of a clustering result with respect to a network snapshot, and the temporal cost calibrates the goodness-of-fit of the current clustering result with respect to either historic data or historic clustering results. Evolutionary clustering aims to minimize the overall cost.

Here, we use spectral clustering as a vehicle to illustrate instantiations of evolutionary clustering. For presentation convenience, we use S_t to denote the community indicator at timestamp t, and L_t the graph Laplacian at time t. Given a network snapshot $A^{(t)}$, the snapshot cost of a community partition according to spectral clustering (as discussed in Section 3.3.4) should be

$$CS_t = Tr(S_t^T L_t S_t), \quad s.t. \quad S_t^T S_t = I_k. \tag{5.2}$$

where $L_t = I - D_t^{-\frac{1}{2}} A^{(t)} D_t^{-\frac{1}{2}}$ is the normalized graph Laplacian for network snapshot $A^{(t)}$, and k is the number of communities.

The next task is to define a proper criterion for temporal cost. The temporal cost can be defined as the difference of current community membership and previous timestamp. Intuitively, one may define the temporal cost as

$$CT_t = \|S_t - S_{t-1}\|^2. \tag{5.3}$$

However, it can be shown that the optimal solution of spectral clustering is not unique. With an orthogonal transformation applied to a solution S, we may obtain another equivalently valid solution. Given an orthogonal matrix $Q \in R^{k \times k}$ such that $Q^T Q = Q Q^T = I_k$, let $\tilde{S} = S Q$. It follows that

$$Tr(\tilde{S}^T L \tilde{S}) = Tr(Q^T S^T L S Q) = Tr(S^T L S Q Q^T) = Tr(S^T L S).$$

Consequently, a solution to the spectral clustering is not unique. Any orthogonal transformation multiplied by the solution is also an equivalent solution. A larger value in Eq. (5.3) does not necessarily indicate a bigger difference between S_t and S_{t-1}. Alternatively, the difference between the two matrix can be defined as

$$CT_t = \frac{1}{2}\|S_t S_t^T - S_{t-1} S_{t-1}^T\|^2. \tag{5.4}$$

Here, $1/2$ is added for derivation convenience. It can be shown (Chi et al., 2007) that

$$Cost_t = Tr\left[S_t^T \tilde{L}_t S_t\right]$$

with

$$\tilde{L}_t = I - \alpha \cdot D_t^{-1/2} A^{(t)} D_t^{-1/2} - (1-\alpha) \cdot S_{t-1} S_{t-1}^T. \tag{5.5}$$

Hence, the optimal solution S_t corresponds to the top minimal eigenvectors (excluding the first one) of a modified graph Laplacian defined in Eq. (5.5).

Consider the example in Figure 5.4. The optimal S_1 based on spectral clustering is listed below:

$$S_1 = \begin{bmatrix} 0.33 & -0.44 \\ 0.27 & -0.43 \\ 0.33 & -0.44 \\ 0.38 & -0.16 \\ 0.38 & 0.24 \\ 0.38 & 0.24 \\ 0.38 & 0.38 \\ 0.33 & 0.30 \\ 0.19 & 0.23 \end{bmatrix}.$$

Let $\alpha = 0.7$. Hence, the modified Laplacian for time T_2 is computed as

$$\tilde{L}_t = \begin{bmatrix} 0.91 & -0.42 & -0.33 & -0.21 & -0.01 & -0.01 & 0.01 & 0.01 & 0.01 \\ -0.42 & 0.92 & -0.08 & -0.27 & 0.00 & 0.00 & 0.02 & 0.01 & 0.01 \\ -0.33 & -0.08 & 0.91 & -0.22 & -0.25 & -0.01 & 0.01 & 0.01 & 0.01 \\ -0.21 & -0.27 & -0.22 & 0.95 & -0.18 & -0.24 & -0.02 & -0.02 & -0.01 \\ -0.01 & 0.00 & -0.25 & -0.18 & 0.94 & -0.06 & -0.37 & -0.06 & -0.04 \\ -0.01 & 0.00 & -0.01 & -0.24 & -0.06 & 0.94 & -0.07 & -0.45 & -0.04 \\ 0.01 & 0.02 & 0.01 & -0.02 & -0.37 & -0.07 & 0.91 & -0.44 & -0.05 \\ 0.01 & 0.01 & 0.01 & -0.02 & -0.06 & -0.45 & -0.44 & 0.94 & -0.04 \\ 0.01 & 0.01 & 0.01 & -0.01 & -0.04 & -0.04 & -0.05 & -0.04 & 0.97 \end{bmatrix}.$$

With this modified graph Laplacian, we can partition nodes into two communities at snapshot T_2: $\{1, 2, 3, 4\}$, and $\{5, 6, 7, 8, 9\}$ following standard spectral clustering. Though node 9 becomes isolated at T_2, evolutionary clustering considers this more like a noise and assigns node 9 to be in the same community as nodes $\{5, 6, 7, 8\}$.

Another way to define temporal cost is to enforce current community membership match with the network of previous timestamp. If the community evolution is smooth, then the clustering result derived from current timestamp t should work well for the network snapshot at timestamp $t - 1$ as well. Thus, the temporal cost can be approximated by

$$CT_t = Tr(S_t^T L_{t-1} S_t) \qquad (5.6)$$

Combining the snapshot cost in Eq. (5.2) and the temporal cost in Eq. (5.6), we have the following cost at timestamp t:

$$
\begin{aligned}
Cost_t &= \alpha \cdot CS_t + (1 - \alpha)] \cdot CT_t \\
&= \alpha \cdot Tr[S_t^T L_t S_t] + (1 - \alpha) \cdot Tr[S_t^T L_{t-1} S_t] \\
&= Tr\left[S_t^T (\alpha \cdot L_t + (1 - \alpha) \cdot L_{t-1}) S_t \right] \\
&= Tr\left[S_t^T \tilde{L}_t S_t \right]
\end{aligned}
$$

with

$$\tilde{L}_t = \alpha \cdot L_t + (1 - \alpha) \cdot L_{t-1} = I - \alpha \cdot D_t^{-1/2} A^{(t)} D_t^{-1/2} - (1 - \alpha) \cdot D_{t-1}^{-1/2} A^{(t-1)} D_{t-1}^{-1/2} \quad (5.7)$$

Clearly, rather than optimizing the community structure with respect to the graph Laplacian L_t, now we optimize with respect to a modified Laplacian defined in Eq. (5.7), It is a weighted combination of L_t and L_{t-1}. This combination can be considered as a smoothing effect of graph Laplacian between consecutive timestamps.

Let us revisit the example in Figure 5.4. The graph Laplacians for T_1 and T_2 are:

$$
L_1 =
\begin{bmatrix}
1.00 & -0.41 & -0.33 & -0.29 & 0 & 0 & 0 & 0 & 0 \\
-0.41 & 1.00 & -0.41 & 0 & 0 & 0 & 0 & 0 & 0 \\
-0.33 & -0.41 & 1.00 & -0.29 & 0 & 0 & 0 & 0 & 0 \\
-0.29 & 0 & -0.29 & 1.00 & -0.25 & -0.25 & 0 & 0 & 0 \\
0 & 0 & 0 & -0.25 & 1.00 & -0.25 & -0.25 & -0.29 & 0 \\
0 & 0 & 0 & -0.25 & -0.25 & 1.00 & -0.25 & -0.29 & 0 \\
0 & 0 & 0 & 0 & -0.25 & -0.25 & 1.00 & -0.29 & -0.50 \\
0 & 0 & 0 & 0 & -0.29 & -0.29 & -0.29 & 1.00 & 0 \\
0 & 0 & 0 & 0 & 0 & 0 & -0.50 & 0 & 1.00
\end{bmatrix}
$$

$$
L_2 =
\begin{bmatrix}
1.00 & -0.41 & -0.33 & -0.26 & 0 & 0 & 0 & 0 & 0 \\
-0.41 & 1.00 & 0 & -0.32 & 0 & 0 & 0 & 0 & 0 \\
-0.33 & 0 & 1.00 & -0.26 & -0.33 & 0 & 0 & 0 & 0 \\
-0.26 & -0.32 & -0.26 & 1.00 & -0.26 & -0.32 & 0 & 0 & 0 \\
0 & 0 & -0.33 & -0.26 & 1.00 & 0 & -0.41 & 0 & 0 \\
0 & 0 & 0 & -0.32 & 0 & 1.00 & 0 & -0.50 & 0 \\
0 & 0 & 0 & 0 & -0.41 & 0 & 1.00 & -0.50 & 0 \\
0 & 0 & 0 & 0 & 0 & -0.50 & -0.50 & 1.00 & 0 \\
0 & 0 & 0 & 0 & 0 & 0 & 0 & 0 & 0
\end{bmatrix}
$$

If we set $\alpha = 0.7$. we have the following smoothed graph Laplacian for time T_2.

$$\widetilde{L}_2 = \begin{bmatrix} 1.00 & -0.41 & -0.33 & -0.27 & 0 & 0 & 0 & 0 & 0 \\ -0.41 & 1.00 & -0.12 & -0.22 & 0 & 0 & 0 & 0 & 0 \\ -0.33 & -0.12 & 1.00 & -0.27 & -0.23 & 0 & 0 & 0 & 0 \\ -0.27 & -0.22 & -0.27 & 1.00 & -0.25 & -0.30 & 0 & 0 & 0 \\ 0 & 0 & -0.23 & -0.26 & 1.00 & -0.08 & -0.36 & -0.09 & 0 \\ 0 & 0 & 0 & -0.30 & -0.08 & 1.00 & -0.08 & -0.44 & 0 \\ 0 & 0 & 0 & 0 & -0.36 & -0.08 & 1.00 & -0.44 & -0.15 \\ 0 & 0 & 0 & 0 & -0.09 & -0.44 & -0.44 & 1.00 & 0 \\ 0 & 0 & 0 & 0 & 0 & 0 & -0.15 & 0 & 0.30 \end{bmatrix}$$

Based on this graph Laplacian, we can follow standard spectral clustering as we discussed in Section 3.3.4 to find two communities $\{1, 2, 3, 4\}$ and $\{5, 6, 7, 8, 9\}$.

In summary, for evolutionary spectral clustering, we modify the graph Laplacian by incorporating the information from clustering results (Eq. (5.5) in the previous timestamp, or the network snapshot (Eq. (5.7)). A similar evolutionary clustering scheme has been instantiated with different community detection methods, including latent space models (Sarkar and Moore, 2005), probabilistic soft clustering (Lin et al., 2009), block model approximation (Yang et al., 2009), graph coloring (Tantipathananandh et al., 2007), and a particle-and-density method (Kim and Han, 2009). In reality, the number of communities can change as new nodes and connections might be added. However, evolutionary clustering assumes the number of communities remains unchanged. It requires additional effort to handle these cases.

5.1.3   SEGMENT-BASED CLUSTERING WITH EVOLVING NETWORKS

While independent clustering at each network snapshot may output specious evolution patterns, the evolutionary clustering, by taking temporal information of networks, enforces smoothness between detected communities at consecutive timestamps. This assumption, however, could result in the failure of capturing drastic changes due to external events. Hence, one needs to balance between gradual change under normal circumstances and drastic changes caused by major events.

Another assumption that has been employed is that the community structure remains unchanged in a segment of time. Suppose a sequence of p network snapshots, $\{A^{(1)}, \cdots, A^{(p)}\}$, we can cut it into several segments, say $\{A^{(1)}, \cdots, A^{(q_1)}\}$ is one segment, $A^{(q_1+1)}, A^{(q_2+2)}, \cdots, A^{(q_2)}\}$ is another segment, and so on. In each network segment, the community structure remains a constant. Between two consecutive segments, there is a time point signaling the change. GraphScope (Sun et al., 2007) is such a graph clustering algorithm for segment-based community detection. Essentially, if the network connections do not change much over time, consecutive network snapshots should have the same community structure, thus should be grouped together into one segment. Whenever a new network snapshot does not fit into an existing segment (say, the community structure of current

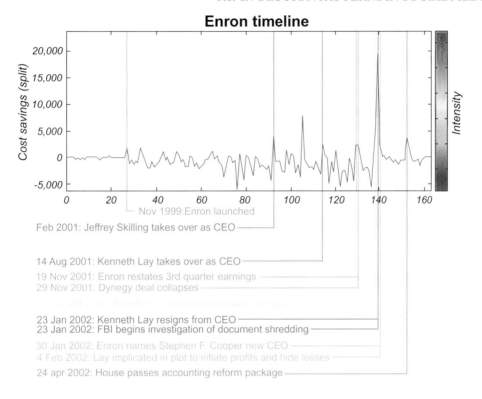

Figure 5.5: Change Points in Enron Email Communication (based on (Sun et al., 2007))

segment induces a high cost on this new snapshot), GraphScope introduces a change point and starts a new segment.

Within one network segment (i.e., multiple consecutive network snapshots), the community partition does not change. A cost function can be defined given a network segment and its partition assignment. GraphScope (Sun et al., 2007), for example, defines the cost as the minimum number of bits in order to compress the network interaction and transmit the information hidden in the partition of a segment. Whenever a new network snapshot arrives, it computes the cost of adding the new snapshot to the old segment, as well as the cost of treating the snapshot as a new segment. If the former cost is lower, then the new snapshot is combined with the old segment and its corresponding community partition is updated; otherwise, a new segment should start, indicating a change point. Figure 5.5 shows those change points found by GraphScope on the Enron email communication network. The X-axis denotes time, each week from January 1999 to July 2002. The Y-axis shows the cost saving of starting a new segment compared with that of incorporating the network snapshot into current segment. A large cost indicates a change point. It is observed that most of the change points coincides with some significant external events.

Identifying community evolution is still a fledgling field. All the methods discussed so far in Section 5.1 require an input of a series of network snapshots. But in reality, a dynamic network is presented in a stream of pairwise interactions. Sometimes, it is not clear what should be a proper time interval to construct network snapshots. It can be tricky to break down a stream of interactions into different snapshots. Distinctive patterns might be found with varying time scales. Novel algorithms are to be developed to discover evolving communities in a network stream.

5.2   CLASSIFICATION WITH NETWORK DATA

In a network environment, individual behaviors are often correlated. For example, if one's friends buy something, there is a better-than-average chance that she will buy it, too. Such correlation between user behaviors can be attributed to both *influence* and *homophily* as discussed in Chapter 2. When an individual is exposed to a social environment, their behavior or decisions are likely to be influenced by each other. This influence naturally leads to correlation between connected users. Since we tend to link up with one another in ways that confirm rather than test our core beliefs, we are more likely to connect to others sharing certain similarities. Consequently, connected users demonstrate correlations in terms of behavior.

An intriguing question is whether one can leverage extant correlations in social media networks to infer user behavior. Take social networking advertising as an example. A common approach to targeted marketing in the past is to build a model mapping from user profiles (e.g., geography location, education level, gender) or user online activities (e.g., search queries, clicked web pages) to ads categories. Since social media often comes with a friendship network between users and their daily interactions, how can we exploit this rich interaction information to infer the ads that might attract a user? Or more generally, how to integrate both user attribute information with their interaction information together for more effective targeting?

These problems can be formulated as a classification task with network data. In particular, the problem of learning with network data can be stated as follows:

> Given a social network with behavior information of some actors, and individual attributes if possible, how can we infer the behavior outcome of the remaining ones within the same network?

Typically, the behavior information can be represented as a label, e.g., whether a user clicks on an ad, whether he would like to join a group of interest, etc. Figure 5.6 shows one example of classification with network data. In the network, four nodes are provided with label information. Nodes 6 and 7 are labeled as positive, while nodes 1 and 2 as negative. Each node could have additional information such as profile attributes. The goal is to predict labels for other five nodes within the same network.

Classification with network data is also known as *within-network classification*, *graph-based classification*, or a special case of *relational learning*. Of course, such a classification problem is not confined to only social media networks. Other instantiations include classification based on tradi-

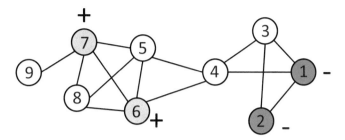

Figure 5.6: An Example for Classification with Network Data

tional web with hyperlink information, email communication networks, and mobile call networks as well.

5.2.1   COLLECTIVE CLASSIFICATION

When data instances are presented in a network form, the samples are not independently and identically distributed (i.i.d.). Their class labels tend to be interdependent on each other. Therefore, collective classification (Sen et al., 2008) is proposed to take into account this autocorrelation present in labels of adjacent nodes. Let variable y_i denote the label of node v_i. And $y_i = 1$ if node v_i is positive, and $y_i = 0$ if negative. A commonly used Markov assumption is that the label of one node depends on the labels (and attributes) of its neighbors. Specifically,

$$p(y_i|A) = p(y_i|N_i)$$

Here, A is the network, and N_i denotes the neighborhood of node v_i. For example, in the network in Figure 5.6, in order to determine the label for node 5, we only need to collect information about nodes 4, 6, 7, and 8.

In order to address the classification task with network data, we typically need three components (Macskassy and Provost, 2007): a *local classifier*, a *relational classifier* and *collective inference*. A general collective classification process can be decomposed into the following steps:

1. Build a *local classifier* to predict labels based on node attributes. Local classifiers are models to map from one node's attributes to the label of the node. Local classifiers do not employ any network information. It only considers the attribute information of nodes in a conventional sense of classification. Hence, conventional classifiers such as naïve Bayes classifiers, logistic regression and support vector machines (Tan et al., 2005) can be used. This local classifier can generate priors that comprise the initial state for subsequent learning and inference.

2. Take the prediction of local classifier as initial label assignment, build a *relational classifier* based on labeled nodes and their neighboring labels. A relational classifier makes use of connections as well as the values of the attributes of related entities. This relational classifier typically takes

the labels (and attributes) of one's neighboring nodes as input. If no attributes are available, then only the labels of one node's neighbors are taken as input. For instance, we may take the percentage of positive and negative nodes in neighborhood as feature input to build a relational classifier.

3. Perform *collective inference* by applying the relational classifier to nodes in the network iteratively. Collective inference determines the labels of all nodes in one network as a whole, rather than independently predicting the label of each node. It aims to find an assignment of labels such that the inconsistency between neighboring nodes are minimized. It iteratively updates the label (a.k.a. *iterative classification*) or class membership probability (a.k.a. *relaxation labeling*) of one node given its neighbors by applying the constructed relational classifier in Step 2.

One recommended approach to handle classification with network data is weighted vote relational neighborhood (wvRN) classifier in conjunction with relaxation labeling (Macskassy and Provost, 2007). Take Figure 5.6 as an example. Suppose we only have access to the network connectivity information between nodes and to four labeled nodes as indicated in the figure. Since there is no information about node attributes, no local classifier is constructed. Instead, we build a relational classifier directly. The wvRN classifier builds a model as follows:

$$
\begin{aligned}
p(y_i = 1|N_i) &= \frac{1}{\sum_{v_j \in N_i} A_{ij}} \sum_{v_j \in N_i} A_{ij} \cdot p(y_j = 1|N_j) \\
&= \frac{1}{|N_i|} \sum_{v_j \in N_i} p(y_j = 1|N_j).
\end{aligned}
\tag{5.8}
$$

It is simply the weighted average of the class memberships of neighboring nodes. Clearly, such a classifier does not require any learning during training.

As for prediction, *collective inference* is exploited to take into account the interdependency between the labels of neighboring nodes. As for the example in Figure 5.6, let's initiate all the unlabeled nodes with $p(y_i = 1|N_i) = 0.5$[3]. Then we can iteratively update the class membership of each node by applying the aforementioned wvRN classifier. Suppose we update the class membership of unlabeled nodes following their node identifications, i.e., update nodes following the order $3, 4, 5, 8,$ and 9. For node 3, it has three friends (nodes $1, 2$ and 4), with $p(y_1 = 1|N_1) = 0, p(y_2 = 1|N_2) = 0,$ $p(y_4 = 1|N_4) = 0.5$. Hence, its class membership is updated as

$$
p(y_3 = 1|N_3) = \frac{1}{3}(0 + 0 + 0.5) = 0.17.
$$

---

[3]This probability can be established by a local classifier in Step 1 of collective classification.

In a similar vein, we can compute nodes 4 and 5' class memberships as follows:

$$p(y_4 = 1|N_4) = \frac{1}{4}\{p(y_1 = 1|N_1) + p(y_3 = 1|N_3) + p(y_5 = 1|N_5) + p(y_6 = 1|N_6)\}$$

$$= \frac{1}{4}(0 + 0.17 + 0.5 + 1) = 0.42.$$

$$p(y_5 = 1|N_5) = \frac{1}{4}\{p(y_4 = 1|N_4) + p(y_6 = 1|N_6) + p(y_7 = 1|N_7) + p(y_8 = 1|N_8)\}$$

$$= \frac{1}{4}(0.42 + 1 + 1 + 0.5) = 0.73.$$

Hence, the class membership of each node is updated one by one. If $\mathbf{p}^{(t)}$ denotes the class memberships at iteration t, it follows that

$$\mathbf{p}^{(1)} = \begin{bmatrix} 0 \\ 0 \\ 0.17 \\ 0.42 \\ 0.73 \\ 1 \\ 1 \\ 0.91 \\ 1.00 \end{bmatrix}, \mathbf{p}^{(2)} = \begin{bmatrix} 0 \\ 0 \\ 0.14 \\ 0.47 \\ 0.85 \\ 1 \\ 1 \\ 0.95 \\ 1.00 \end{bmatrix}, \mathbf{p}^{(3)} = \begin{bmatrix} 0 \\ 0 \\ 0.16 \\ 0.50 \\ 0.86 \\ 1 \\ 1 \\ 0.95 \\ 1.00 \end{bmatrix}, \mathbf{p}^{(4)} = \begin{bmatrix} 0 \\ 0 \\ 0.17 \\ 0.51 \\ 0.87 \\ 1 \\ 1 \\ 0.96 \\ 1.00 \end{bmatrix}, \mathbf{p}^{(5)} = \begin{bmatrix} 0 \\ 0 \\ 0.17 \\ 0.51 \\ 0.87 \\ 1 \\ 1 \\ 0.96 \\ 1.00 \end{bmatrix}.$$

Clearly, after 5 iterations, the class membership stabilizes. As can be seen from this simple example, it requires multiple scans of the whole network to reach convergence for collective inference. Of course, other heuristics can be used to expedite the process, such as shuffling the processing order of nodes at each interaction.

Roughly speaking, the label of one node is determined the distance of the node to other labeled nodes. For instance, node 5 is closer to positive nodes 6 and 7. That is why node 5 has a higher probability of belonging to class +, indicated by $p(y_5 = 1|N_5) = 0.87$ in $\mathbf{p}^{(5)}$. By contrast, node 4 lies in the middle between both positive and negative nodes. Hence, it does not show a strong inclination toward either class, as indicated by $p(y_4 = 1|N_4) = 0.51$. This simple wvRN classifier is highly related to a graph-based semi-supervised learning method following the Gaussian field (Zhu et al., 2003). Alternatively, the collective inference process can be considered as an approximation to label propagation, i.e., labeled nodes propagate their labels to their friends, and then friends of friends, and so on.

5.2.2   COMMUNITY-BASED LEARNING

Connections in social media are often inhomogeneous. The *heterogeneity* present in network connections can hinder the success of collective inference. People can connect to their relatives, colleagues,

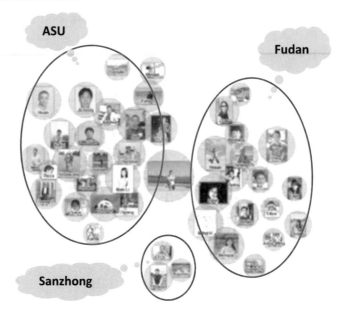

Figure 5.7: An example of a user's contacts in Facebook, involving three different relations: friends met at ASU, undergraduate classmates at Fudan University, and some high school friends at Sanzhong.

college classmates, or some buddies met online. These relations play different roles in helping determine targeted behaviors. For instance, the Facebook contacts of the first author can be seen in three key groups, as shown in Figure 5.7: friends at Arizona State University (ASU), undergraduate classmates at Fudan University, and some high-school friends in Sanzhong. For example, it is reasonable to infer that his friends at ASU are presumably interested to know if Lei is watching an ASU football game, while his other friends at Fudan and Sanzhong are probably indifferent to his excitement[4]. In other words, users can be involved in different relations, or a social network can consist of heterogeneous relations. Therefore, it is not appropriate to directly apply collective inference to this kind of networks as it does not differentiate these heterogeneous relations. It is recommended to differentiate relations for collective classification.

However, detailed relation information is not readily available. Though some sites like LinkedIn and Facebook do ask people how they know each other when they become connected, most of the time, people decline, or simply are too lazy to share such specific information, resulting in a social network among users without explicit information about pairwise relations. Differentiating pairwise relations based on network connectivity alone is by no means an easy task.

Alternatively, we can look at *social dimensions* (Tang and Liu, 2009b) of actors. Social dimensions are latent dimensions defined by social connections. It is introduced to represent latent relations

[4]People in China mainly follow soccer matches.

Table 5.1: Social Dimensions

Actors	ASU	Fudan	Sanzhong
Lei	1	1	1
Actor$_1$	1	0	0
⋮	⋮	⋮	⋱

associated with actors, with each dimension denoting one relation. Suppose two actors a_i and a_j are connected because of relationship R, both a_i and a_j should have a non-zero entry in the social dimension which represents R. Let us revisit the Facebook example. The relations between Lei and his friends can be characterized by three affiliations: Arizona State University (ASU), Fudan University (Fudan), and a high school (Sanzhong). The corresponding social dimensions of actors are shown in Table 5.1. In the table, if one actor belongs to one affiliation, then the corresponding entry is non-zero. Since Lei is a student at ASU, his social dimension includes a non-zero entry for the ASU dimension.

Social dimensions capture prominent interaction patterns presented in a network. Note that one actor is very likely to be involved in multiple different social dimensions (e.g., Lei engages in three different relations in the example). This is consistent with the multi-facet nature of human social life which states that one is likely to be involved in distinctive relations with different people.

The social dimensions shown in Table 5.1 are constructed based on explicit relation information. In reality, without knowing the true relationship, how can we extract social dimensions? One key observation is that actors of the same relation tend to connect to each other as well. For instance, most of Lei's friends at ASU are themselves friends as well. Hence, to infer a social dimension, we first find out a group of people who interact with each other more frequently than random. This boils down to a classical community detection problem. A requirement is that *one actor is allowed to be assigned to multiple communities*. We can resort to soft clustering or methods finding overlapping communities to extract social dimensions.

After we extract the social dimensions, we consider them as normal features and combine them with the behavioral information for supervised learning to construct a classifier. This supervised learning is critical as it will determine which dimensions are relevant to a target behavior and assign proper weights to different social dimensions. In summary, this social-dimension based learning framework (SocioDim) (Tang and Liu, 2010a) consists of three steps:

1. Extract meaningful social dimensions based on network connectivity via community detection.

2. Determine relevant social dimensions (plus node attributes) through supervised learning.

3. Apply the constructed model in Step 2 to the social dimensions of actors without behavioral information, to obtain behavior predictions.

This SocioDim framework basically assumes that an actor's community membership determines his behavior. This can be visualized more clearly using an example in Figure 5.8. The circles in

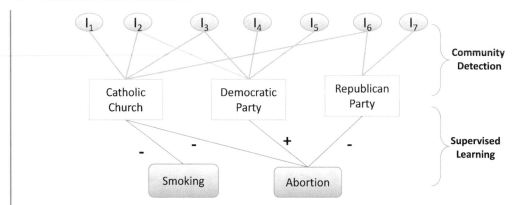

Figure 5.8: A Collective Behavior Model for the SocioDim framework (based on (Tang and Liu, 2010a))

orange denote individuals, the green rectangles affiliations and the red blocks at the bottom behaviors. Individuals are associated with different affiliations in varying degrees and distinctive affiliations regulate the member behavior differently. For instance, in the green blocks, Catholic Church opposes smoking and abortion while Democratic Party supports abortion. Note that some affiliations might have no influence over certain behavior such as the blocks of the Democratic Party and Republican Party over smoking. The final output of individual behavior depends on the affiliation regularization and individual associations. The first step of our proposed SocioDim framework essentially finds out the individual associations and the second step learns the affiliation regularization by assigning weights to different affiliations.

Such a SocioDim learning strategy can be applied to the network in Figure 5.6. We can use community detection methods discussed in Chapter 3 to extract heterogeneous relations in a network. Since one is likely to involve in multiple relations, we resort to *soft clustering* or methods finding *overlapping communities*. The soft community indicator is obtained by spectral clustering with normalized graph Laplacian:

$$S = \begin{bmatrix} 0.33 & -0.44 \\ 0.27 & -0.43 \\ 0.33 & -0.44 \\ 0.38 & -0.16 \\ 0.38 & 0.24 \\ 0.38 & 0.24 \\ 0.38 & 0.38 \\ 0.33 & 0.30 \\ 0.19 & 0.23 \end{bmatrix}.$$

By treating the above soft community indicator as features for each node, we can construct conventional attribute data as in Table 5.2. Based on the community membership (S_1 and S_2), we can

Node	S_1	S_2	Label	Pred. Score	Pred.
1	0.33	-0.44	—		
2	0.27	-0.43	—		
3	0.33	-0.44	?	-0.53	-
4	0.38	-0.16	?	-0.13	-
5	0.38	0.24	?	0.42	+
6	0.38	0.24	+		
7	0.38	0.38	+		
8	0.33	0.30	?	0.50	+
9	0.19	0.23	?	0.38	+

Table 5.2: Communities are Features

construct a classifier such as logistic regression, support vectors machines, decision trees, and many more. A support vector machine can be learned from labeled data, resulting in the following decision function:

$$y = sign(0.16S_1 + 1.39S_2 + 0.03).$$

Applying the classifier to the remaining unlabeled nodes, we have prediction scores and corresponding predictions as in the table.

While such a soft clustering scheme leads to an accurate prediction, it might require extensive memory to hold the community membership. The soft community indicator is a full matrix. When a network scales to millions of nodes, the memory footprint of holding the social dimensions alone becomes unbearable. By contrast, overlapping communities might be able to induce sparse representations. For the example, we can find two overlapping communities $\{1, 2, 3, 4\}$ and $\{4, 5, 6, 7, 8\}$ via the clique percolation method (CPM) (Section 3.1.1). Then its community membership representation would be sparse as follows:

$$
S = \begin{bmatrix}
1 & 0 \\
1 & 0 \\
1 & 0 \\
1 & 1 \\
0 & 1 \\
0 & 1 \\
0 & 1 \\
0 & 1 \\
0 & 0
\end{bmatrix}.
$$

With a sparse representation, the subsequent learning could be accomplished efficiently. However, finding overlapping communities is not a simple task. The CPM method is computational costly. Recently, new techniques to find overlapping communities such as EdgeCluster (Tang and Liu,

Table 5.3: Collective Classification vs. Community-Based Learning		
	Collective Classification	**SocioDim**
Computational cost for training	low	high
Computational cost for prediction	high	low
Handling heterogeneous relations		✓
Handling evolving networks	✓	
Integrating network information and node attributes	✓	✓

2009a) and MROC (Tang et al., 2010) were developed to handle large-scale networks of millions of nodes for classification.

5.2.3 SUMMARY

Collective classification and community-based learning are two very different schemes for classification. Collective classification relies on collective inference to handle the interdependency between connected nodes. It is typically very fast during the training phase, e.g., the wvRN classifier does not require any learning during training. However, the prediction phase can be time-consuming because we need to iteratively update the class or class membership probability for each node. Most collective classification algorithms can be easily adapted to handle an evolving network.

Community-based learning, however, focuses on extracting social dimensions before supervised learning. The computational cost of extracting communities could be expensive, but the prediction using social dimensions is very fast as no collective inference is necessary. Since SocioDim can differentiate connections in a network, thus it typically outperforms collective classification when a network presents heterogeneous relations. However, there is no natural efficient solution for SocioDim to handle an evolving network. Their strengths and weaknesses are shown in Table 5.3.

As mentioned earlier, one may need to integrate network information and node attributes for learning. Collective classification embeds this integration in its relational classifier construction and collective inference while SocioDim separates these two types of information. Communities are extracted from a network and treated as features of nodes, which can be combined with other features (e.g., attribute values) for learning.

Community detection and mining in social media is an exciting and vibrant area with many open research questions. The journey of searching for solutions is full of challenges and uncertainty. We hope this lecture will help make the journey less circuitous and more enjoyable.

APPENDIX A

Data Collection

The first step in many social computing tasks is to collect data. Network data collection in social media is radically different from that in traditional social sciences. Traditional social sciences use surveys and involve subjects in the data collection process. The subject are asked to answer questions about their friends and other related information. A network is then constructed conditioned on the responses. Limited by this process, data collected is of small sizes, typically dozens or hundreds of subjects in one study. By contrast, hundreds and thousands of users of social media produce inordinate amounts of data with rich user interactions, but few of them would be interested to answer any survey questionnaires. Thus, a different approach is needed.

From a social media host's viewpoint, it is straightforward to collect network information. Every piece of data can be stored in databases. The host can query the databases to obtain desired network information. Under some circumstances, researchers may be able to access to the data with an agreement of use. For example, all the participants of the first Netflix prize competition could access the user-movie rating data. Some benchmark network data may also be accessible for research purposes, e.g., http://snap.stanford.edu/data/index.html.

More often than not, a researcher may have to resort to Web crawling to collect network data. There are two common approaches: 1) crawling using site provided APIs, or 2) scraping needed information from rendered html pages. Many social media sites provide APIs, e.g., Twitter[1], Facebook[2], YouTube[3], Flickr[4], Digg[5], etc. Normally, APIs are faster in response to html requests, and the results are provided in a site-specific (semi) structured format. However, restrictions are imposed. Some sites return only the top-k results. One can obtain at most 100 friends per actor through Twitter API. To avoid being overwhelmed by the requests from an IP address, most sites also cap the number of requests from the same crawler token or the same IP address in some time period (say, 100 requests per hour or per day). Twitter allows for 150 requests per hour, for instance. However, if one is registered in Twitter's whitelist, he is allowed to send $20,000$ requests per hour per IP.

One could also parse the available html pages of each user and extract desired information. One can scrutinize the profile pages of users and analyze the templates of html pages to extract related information like friends, location, birthday, etc. Hence, an html parser needs to be developed and customized for each site. It can get most information a user shares online, but requires laborious

[1]http://apiwiki.twitter.com/
[2]http://developers.facebook.com/
[3]http://code.google.com/apis/youtube/overview.html
[4]http://www.flickr.com/services/api/
[5]http://digg.com/api/docs/overview

efforts like parser designing than the crawling using APIs. Parsing html pages is also typically slower than using site provided APIs.

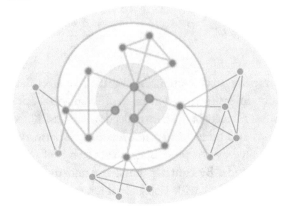

Figure A.1: Snowball Sampling. Blue nodes are the seeds, red and green ones are the nodes in the first and second waves, respectively.

In order to obtain the network data, a standard approach is snowball sampling (Goodman, 1961). It is originally proposed in social sciences to reach those hidden populations that are normally difficult to reach by random sampling, such as drug users, sex workers, and unregulated workers. Snowball sampling allows the sampled individuals to provide information about their contacts. In social media, a similar procedure can be adopted. Figure A.1 shows the idea of snowball sampling. That is, through seed users (blue nodes in the center), we can obtain their friends (in pink), or the first wave. We can then get friends of their friends (in green), the second wave. Essentially, given the nodes in the k-th wave, we expand the network by finding their friends, forming the $(k + 1)$-th wave. In the process, the "snowball" (network of users) is rolling larger and larger. Typically, a breadth-first search procedure is preferred to reduce the sample sensitivity to the initial nodes (Menczer, 2006) because those very popular nodes can be reached in a few hops even starting from unpopular seeds, which enables the collection of sufficient amounts of data. Snowball sampling explores connections starting from seed nodes. Thus, the crawled network is within the same component. Alternative solutions should be explored if one is interested in networks of different components.

Another approach to network data collection is to examine the patterns in user IDs or profile URLs. Flickr, for example, always assigns a user ID in a specified format. Below are three user IDs randomly scraped from Flickr:

30720482@N05
81544085@N00
14048444@N08

The URL template of their corresponding user profile pages looks like http://www.flickr.com/people/[userid]. By changing the first 8 digits and the last one digit, one can randomly generate user IDs, either valid or invalid, and then crawl their corresponding contacts in Flickr if the IDs exist in Flickr. This approach is able to reach nodes that are distributed in different components, whereas snowball crawling can only reach nodes within the same components of the seed nodes. Unfortunately, not all social media sites provide observable patterns as Flickr does. Hence, snowball sampling and its variants are more popular in practice.

APPENDIX B

Computing Betweenness

Betweenness can be defined for both nodes and edges. Node (edge) betweenness is defined as the number of shortest paths that pass one node (or along one edge). Both node betweenness and edge betweenness are commonly used in network analysis. For example, *node betweenness* is adopted for centrality analysis in Section 2.1, and *edge betweenness* is utilized to find the "weakest" tie for hierarchical clustering in Section 3.4.1. We present an algorithm to compute node and edge betweenness in an unweighted undirected network. The algorithm requires $O(n + m)$ space and $O(nm)$ time where n and m are the number of nodes and edges, respectively (Brandes, 2001). With minor changes, it can be adapted to handle directed and/or weighted networks (Brandes, 2001; Newman and Girvan, 2004).

The algorithm traverses each node, finds all shortest paths starting from a source node s, and accumulates betweenness for related nodes and edges. For presentation convenience, the following concepts are defined.

- $g_s[t]$: the geodesic distance from node s to node t.

- $\sigma_s[t]$: the number of shortest paths from node s to node t;

- The *dependency* of shortest paths from node s over node v:

$$\delta_s(v) = \sum_{t \in V} \frac{\sigma_{st}(v)}{\sigma_s[t]}.$$

where $\sigma_{st}(v)$ is the number of shortest paths from s to t that pass node v.

- $P_s[w]$: a *list* of parent nodes of w in shortest paths starting from node s. A node v is a *parent node* of w if $g_s[w] = g_s[v] + 1$. In other words, there is a shortest path starting from node s that has to pass node v in order to reach w.

The pseudocode of the algorithm to compute betweenness is detailed in Figure B.1. Starting from a source node s, the algorithm carries out the following two steps:

1. Conduct a breadth-first search starting from node s in order to determine the number of shortest paths starting from node s and terminating at another node (line $10 - 23$). Once we reach a new node (line $14 - 17$) during the search, its geodesic distance to the source node s is updated. When we find a parent node v for w (line $18 - 21$), the total number of shortest paths reaching w is increased by σ_v as all the shortest paths reaching node v are also shortest paths to w following edge $e(v, w)$.

1: $C_B[v] \leftarrow 0, v \in V$; //node betweenness
2: $C_B[e] \leftarrow 0, e \in E$; //edge betweenness
3: **for** $s \in V$ **do**
4:      $S \leftarrow$ empty stack;
5:      $P_s[w] \leftarrow$ empty list, $w \in V$; //initialize the parent nodes of each node
6:      $g_s[t] = -1$, $t \in V$; $g_s[s] = 0$; //the geodesic distance of each node to s
7:      $\sigma_s[t] = 0$, $t \in V$; $\sigma_s[s] = 1$; //the number of shortest paths from s to t
8:      $Q \leftarrow$ empty queue; //a queue for breadth-first search
9:      enqueue $v_s \rightarrow Q$;
10:      **while** Q not empty **do**
11:          dequeue $v \leftarrow Q$;
12:          push $v \rightarrow S$;
13:          **foreach** neighbor w of v **do**
14:              **if** $g_s[w] < 0$ **then** //encounter w for the first time
15:                  $g_s[w] \leftarrow g_s[v] + 1$;
16:                  enqueue $w \rightarrow Q$;
17:              **end**
18:              **if** $g_s[w] = g_s[v] + 1$ **then** //v is a parent node of w
19:                  $\sigma_s[w] \leftarrow \sigma_s[w] + \sigma_s[v]$;
20:                  append $v \rightarrow P_s[w]$;
21:              **end**
22:          **end**
23:      **end**
24:      $\delta_s[v] = 0, v \in V$;
25:      //S returns vertices in order of non-increasing distance from s
26:      **while** S not empty **do**
27:          pop $w \leftarrow S$;
28:          **for** $v \in P_s[w]$ **do**
29:              $C_B(e(v, w)) \leftarrow C_B(e(v, w)) + \frac{\sigma_s[v]}{\sigma_s[w]} \cdot (1 + \delta_s[w])$;
30:              $\delta_s[v] \leftarrow \delta_s[v] + \frac{\sigma_s[v]}{\sigma_s[w]} \cdot (1 + \delta_s[w])$;
31:          **end**
32:          **if** $w \neq s$, **then** $C_B(w) \leftarrow C_B(w) + \delta_s(w)$;
33:      **end**
34: **end**
35: $C_B[v] \leftarrow C_B[v]/2, v \in V$; // reduce betweenness by half in undirected networks
36: $C_B[e] \leftarrow C_B[e]/2, e \in E$; // as each shortest path is counted twice

Figure B.1: Algorithm to Compute Betweenness in Undirected Unweighted Networks

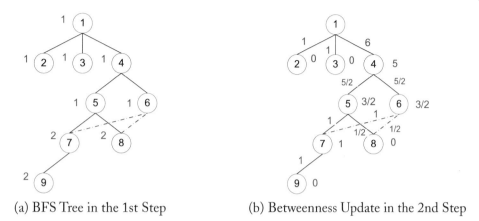

(a) BFS Tree in the 1st Step   (b) Betweenness Update in the 2nd Step

Figure B.2: Algorithm Applied to the Network in Figure 1.1 Starting From Node 1

2. Traverse nodes in the order of non-increasing distance from node s and update the betweenness accordingly (line $26 - 33$). In particular (line 29), given a node w and its parent node v, we update the betweenness of the edge $e(v, w)$ by adding a weight

$$\Delta = \frac{\sigma_s[v]}{\sigma_s[w]} \cdot (1 + \delta_s[w]), \tag{B.1}$$

where $\delta_s[w]$ is the dependency over node w, i.e., the weighted sum of shortest paths passing node w. Meanwhile, there is a total weight of 1 for the shortest paths terminating at node w. Among all these shortest paths that terminate at or pass node w, only $\frac{\sigma_s[v]}{\sigma_s[w]}$ of them are through node w's parent node v. The weight in Eq. (B.1) is added to the parent node v's betweenness as well (line 30).

Take the network in Figure 1.1 as an example. Suppose we start with node 1 to perform a breadth-first search. The corresponding BFS tree in the first step is shown in Figure B.2(a) where the number besides each node denotes $\sigma_1(v)$. In the figure, the dashed lines denote the cases when we encounter a parent node as in line $18 - 21$. That's why node 7, 8 and 9 have $\sigma_1(v) = 2$. Then in the second step, we traverse all nodes in a bottom-up fashion. The corresponding weights updated to edges and nodes are shown in Figure B.2(b). In the figure, the blue numbers besides each edge denote betweenness update Δ to each edge, and the red numbers besides each node are the node's dependency $\delta_1(v)$. For example, we start from node 9 with $\delta_1(9) = 0$. Since node 7 is a parent node of node 9, the weight update on edge $e(7,9)$ should be $\frac{\sigma_1(7)}{\sigma_1(9)}(1 + \delta_1(9)) = \frac{2}{2}(1 + 0) = 1$. Note that the quantities shown in the figure are different from the betweenness reported in Section 2.1. This is because we only count the shortest paths from node 1 here. After we exhaust shortest paths from all nodes, we should get the same betweenness values for each node as reported.

APPENDIX C

k-Means Clustering

The k-means clustering algorithm has been used frequently to find communities in networks in Chapters 3 and 4. k-means clustering finds clusters of data in attribute format. Given a parameter k, the k-means clustering algorithm partitions data instances into k clusters. Each cluster is associated with a centroid (center point). The centroid of a cluster is typically computed as the mean of the data points assigned to the cluster. And each data point is assigned to the closest cluster, or the distance between its centroid and the data point is the shortest. The closeness of data points to a centroid depends on the definition of distance or the inverse of similarity. In classical k-means, Euclidean distance is employed and a data instance is assigned to the centroid with minimum distance. As in Section 3.3.1, we can also assign a data instance to a centroid with maximum cosine similarity or Jaccard similarity. The basic k-means clustering algorithm (Tan et al., 2005) is listed in Figure C.1.

1: select k points as initial centroids;
2: **repeat**
3:     form k clusters by assigning all points to their closest centroids;
4:     recompute the centroid of each cluster;
5: **until** the centroids do not change.

Figure C.1: The k-Means Clustering Algorithm

Here, we apply k-means clustering algorithm with $k = 2$ to a data set X of 7 instances listed below:

$$X = \begin{bmatrix} & X_1 & X_2 \\ 1 & -1.51 & 0.06 \\ 2 & -2.56 & 0.17 \\ 3 & -1.51 & 0.06 \\ 4 & -0.53 & -0.01 \\ 5 & 0.47 & -0.08 \\ 6 & 0.47 & -0.08 \\ 7 & 1.47 & 0.14 \\ 8 & 1.29 & -0.95 \\ 9 & 2.42 & 0.70 \end{bmatrix}$$

Each column above indicates one attribute (X_1 and X_2, respectively). We start by randomly selecting two data points (say, #2 and #4) as the initial centroids:

	X_1	X_2
Centroid 1	-2.56	0.17
Centroid 2	-0.53	-0.01

The Euclidean distance of each data instance to the centroid is given below:

	Centroid 1	Centroid 2	Cluster Assignment
1	1.06	**0.98**	2
2	**0**	2.04	1
3	1.06	**0.98**	2
4	2.04	**0**	2
5	3.04	**1.01**	2
6	3.04	**1.01**	2
7	4.03	**2.01**	2
8	4.02	**2.06**	2
9	5.01	**3.04**	2

We assign each data instance to the cluster with the closest centroid. Thus, we have the assignment as in the last column in the above table. The cluster assignment is also shown in the upper-left corner of Figure C.2, with circles and squares each denote one cluster. Based on the cluster assignment, we can compute an updated centroid for each cluster as indicated by the red triangles. Then, we can repeat the above process to assign data points and update centroids for the clusters. The clusters and centroids in each iteration are depicted in Figure C.2. After 4 iterations, the centroids do not change anymore, and the k-means clustering algorithm converges. The two clusters are: {1, 2, 3, 4} and {5, 6, 7, 8, 9}.

Indeed, the data X in the above example is the soft community indicator S (Eq. (3.9)) following the latent space model (Section 3.3.2). Clearly, the final clusters recover the community structure in the network in Figure 1.1.

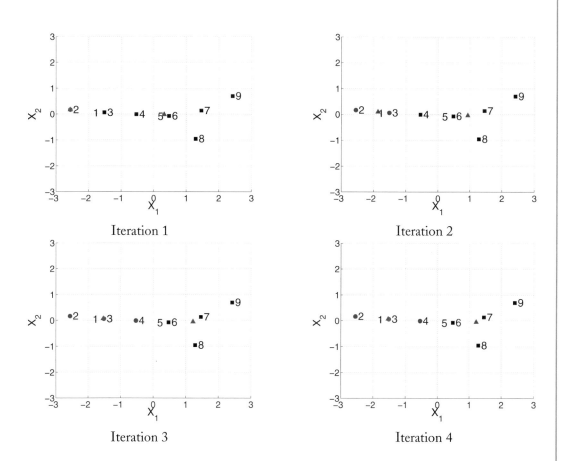

Figure C.2: Cluster Assignment and Corresponding Centroids in Each Iteration. Circles and squares each denote a cluster, and triangles are the centroids.

Bibliography

J. Abello, M. G. C. Resende, and S. Sudarsky. Massive quasi-clique detection. In *LATIN*, pages 598–612, 2002. DOI: 10.1007/3-540-45995-2_51 34, 35

N. Agarwal, H. Liu, L. Tang, and P. S. Yu. Identifying the influential bloggers in a community. In *WSDM '08: Proceedings of the international conference on Web search and web data mining*, pages 207–218, New York, NY, USA, 2008. ACM. ISBN 978-1-59593-927-9. DOI: 10.1145/1341531.1341559 8

A. Anagnostopoulos, R. Kumar, and M. Mahdian. Influence and correlation in social networks. In *KDD '08: Proceeding of the 14th ACM SIGKDD international conference on Knowledge discovery and data mining*, pages 7–15, New York, NY, USA, 2008. ACM. ISBN 978-1-60558-193-4. DOI: 10.1145/1401890.1401897 27, 28, 29

R. Andersen and K. J. Lang. Communities from seed sets. In *WWW '06: Proceedings of the 15th international conference on World Wide Web*, pages 223–232, New York, NY, USA, 2006. ACM. ISBN 1-59593-323-9. DOI: 10.1145/1135777.1135814 9

S. Aral, L. Muchnik, and A. Sundararajan. Distinguishing influence-based contagion from homophily-driven diffusion in dynamic networks. *Proceedings of the National Academy of Sciences*, 106(51):21544–21549, 2009. URL http://www.pnas.org/content/106/51/21544.full. DOI: 10.1073/pnas.0908800106 29

S. Asur, S. Parthasarathy, and D. Ucar. An event-based framework for characterizing the evolutionary behavior of interaction graphs. In *KDD '07: Proceedings of the 13th ACM SIGKDD international conference on Knowledge discovery and data mining*, pages 913–921, New York, NY, USA, 2007. ACM. ISBN 978-1-59593-609-7. DOI: 10.1145/1281192.1281290 9, 76, 78

L. Backstrom, D. Huttenlocher, J. Kleinberg, and X. Lan. Group formation in large social networks: membership, growth, and evolution. In *KDD '06: Proceedings of the 12th ACM SIGKDD international conference on Knowledge discovery and data mining*, pages 44–54, New York, NY, USA, 2006. ACM. ISBN 1-59593-339-5. DOI: 10.1145/1150402.1150412 9, 76

L. Backstrom, C. Dwork, and J. Kleinberg. Wherefore art thou r3579x?: anonymized social networks, hidden patterns, and structural steganography. In *WWW '07: Proceedings of the 16th international conference on World Wide Web*, pages 181–190, New York, NY, USA, 2007. ACM. ISBN 978-1-59593-654-7. DOI: 10.1145/1242572.1242598 11

A.-L. Barabási and R. Albert. Emergence of Scaling in Random Networks. *Science*, 286(5439): 509–512, 1999. URL http://www.sciencemag.org/cgi/content/abstract/286/5439/ 509. DOI: 10.1126/science.286.5439.509 7, 8

L. Becchetti, P. Boldi, C. Castillo, and A. Gionis. Efficient semi-streaming algorithms for local triangle counting in massive graphs. In *KDD '08: Proceeding of the 14th ACM SIGKDD international conference on Knowledge discovery and data mining*, pages 16–24, New York, NY, USA, 2008. ACM. ISBN 978-1-60558-193-4. DOI: 10.1145/1401890.1401898 8

V. Blondel, J. Guillaume, R. Lambiotte, and E. Lefebvre. Fast unfolding of communities in large networks. *Journal of Statistical Mechanics: Theory and Experiment*, 2008:P10008, 2008. DOI: 10.1088/1742-5468/2008/10/P10008 48

I. Borg and P. Groenen. *Modern Multidimensional Scaling: theory and applications*. Springer, 2005. 37

U. Brandes. A faster algorithm for betweenness centrality. *Journal of Mathematical Sociology*, 25(2): 163–177, 2001. DOI: 10.1080/0022250X.2001.9990249 16, 18, 46, 48, 97

J. S. Breese, D. Heckerman, and C. Kadie. Empirical analysis of predictive algorithms for collaborative filtering. In *UAI '98: Proceedings of the 14th Conference on Uncertainty in Artificial Intelligence*, 1998. 10

D. Chakrabarti and C. Faloutsos. Graph mining: Laws, generators, and algorithms. *ACM Comput. Surv.*, 38(1):2, 2006. ISSN 0360-0300. 7, 8

D. Chakrabarti, R. Kumar, and A. Tomkins. Evolutionary clustering. In *KDD '06: Proceedings of the 12th ACM SIGKDD international conference on Knowledge discovery and data mining*, pages 554–560, New York, NY, USA, 2006. ACM. ISBN 1-59593-339-5. DOI: 10.1145/1150402.1150467 79

W. Chen, Y. Wang, and S. Yang. Efficient influence maximization in social networks. In *KDD '09: Proceedings of the 15th ACM SIGKDD international conference on Knowledge discovery and data mining*, pages 199–208, New York, NY, USA, 2009. ACM. ISBN 978-1-60558-495-9. DOI: 10.1145/1557019.1557047 26

W. Chen, C. Wang, and Y. Wang. Scalable influence maximization for prevalent viral marketing in large-scale social networks. In *KDD '10: Proceedings of the 16th ACM SIGKDD international conference on Knowledge discovery and data mining*, pages 1029–1038, New York, NY, USA, 2010. ACM. ISBN 978-1-4503-0055-1. DOI: 10.1145/1835804.1835934 26

Y. Chi, X. Song, D. Zhou, K. Hino, and B. L. Tseng. Evolutionary spectral clustering by incorporating temporal smoothness. In *KDD '07: Proceedings of the 13th ACM SIGKDD international conference on Knowledge discovery and data mining*, pages 153–162, New York, NY, USA, 2007. ACM. ISBN 978-1-59593-609-7. DOI: 10.1145/1281192.1281212 79, 80

N. A. Christakis and J. H. Fowler. The spread of obesity in a large social network over 32 years. *The New England Journal of Medicine*, 357(4):370–379, 2007. DOI: 10.1056/NEJMsa066082 29

A. Clauset, M. Newman, and C. Moore. Finding community structure in very large networks. *Arxiv preprint cond-mat/0408187*, 2004. DOI: 10.1103/PhysRevE.70.066111 48, 52

D. Crandall, D. Cosley, D. Huttenlocher, J. Kleinberg, and S. Suri. Feedback effects between similarity and social influence in online communities. In *KDD '08: Proceeding of the 14th ACM SIGKDD international conference on Knowledge discovery and data mining*, pages 160–168, New York, NY, USA, 2008. ACM. ISBN 978-1-60558-193-4. DOI: 10.1145/1401890.1401914 27, 29

P. Desikan and J. Srivastava. I/o efficient computation of first order markov measures for large and evolving graphs. In *Proceedings of the Tenth Workshop on Web Mining and Web Usage Analysis (WebKDD)*, 2008. 8

I. S. Dhillon. Co-clustering documents and words using bipartite spectral graph partitioning. In *KDD '01: Proceedings of the seventh ACM SIGKDD international conference on Knowledge discovery and data mining*, pages 269–274, New York, NY, USA, 2001. ACM. ISBN 1-58113-391-X. DOI: 10.1145/502512.502550 69

J. Diesner, T. L. Frantz, and K. M. Carley. Communication networks from the enron email corpus "it's always about the people. enron is no different". *Comput. Math. Organ. Theory*, 11(3):201–228, 2005. ISSN 1381-298X. DOI: 10.1007/s10588-005-5377-0 6

P. Domingos and M. Richardson. Mining the network value of customers. In *KDD '01: Proceedings of the seventh ACM SIGKDD international conference on Knowledge discovery and data mining*, pages 57–66, New York, NY, USA, 2001. ACM. ISBN 1-58113-391-X. DOI: 10.1145/502512.502525 24

Y. Dourisboure, F. Geraci, and M. Pellegrini. Extraction and classification of dense communities in the web. In *WWW '07: Proceedings of the 16th international conference on World Wide Web*, pages 461–470, New York, NY, USA, 2007. ACM. ISBN 978-1-59593-654-7. DOI: 10.1145/1242572.1242635 9

D. Easley and J. Kleinberg. *Networks, Crowds, and Markets: Reasoning About a Highly Connected World*. Cambridge University Press, 2010. 19, 27

K. El-Arini, G. Veda, D. Shahaf, and C. Guestrin. Turning down the noise in the blogosphere. In *KDD '09: Proceedings of the 15th ACM SIGKDD international conference on Knowledge discovery and data mining*, pages 289–298, New York, NY, USA, 2009. ACM. ISBN 978-1-60558-495-9. DOI: 10.1145/1557019.1557056 24

G. W. Flake, S. Lawrence, and C. L. Giles. Efficient identification of web communities. In *KDD '00: Proceedings of the sixth ACM SIGKDD international conference on Knowledge discovery and data mining*, pages 150–160, New York, NY, USA, 2000. ACM. ISBN 1-58113-233-6. DOI: 10.1145/347090.347121 9, 52

R. W. Floyd. Algorithm 97: Shortest path. *Commun. ACM*, 5(6):345, 1962. ISSN 0001-0782. DOI: 10.1145/367766.368168 18

S. Fortunato. Community detection in graphs. *Physics Reports*, 486(3-5):75 – 174, 2010. ISSN 0370-1573. URL http://www.sciencedirect.com/science/article/B6TVP-4XPYXF1-1/2/99061fac6435db4343b2374d26e64ac1. DOI: 10.1016/j.physrep.2009.11.002 53

D. Gibson, R. Kumar, and A. Tomkins. Discovering large dense subgraphs in massive graphs. In *VLDB '05: Proceedings of the 31st international conference on Very large data bases*, pages 721–732. VLDB Endowment, 2005. ISBN 1-59593-154-6. 9, 36

E. Gilbert and K. Karahalios. Predicting tie strength with social media. In *CHI '09: Proceedings of the 27th international conference on Human factors in computing systems*, pages 211–220, New York, NY, USA, 2009. ACM. ISBN 978-1-60558-246-7. DOI: 10.1145/1518701.1518736 18, 20

G. H. Golub and C. F. Van Loan. *Matrix computations (3rd ed.)*. Johns Hopkins University Press, Baltimore, MD, USA, 1996. ISBN 0-8018-5414-8. 18, 71

L. Goodman. Snowball sampling. *The Annals of Mathematical Statistics*, 32(1):148–170, 1961. DOI: 10.1214/aoms/1177705148 94

A. Goyal, F. Bonchi, and L. V. Lakshmanan. Learning influence probabilities in social networks. In *WSDM '10: Proceedings of the third ACM international conference on Web search and data mining*, pages 241–250, New York, NY, USA, 2010. ACM. ISBN 978-1-60558-889-6. DOI: 10.1145/1718487.1718518 21

M. Granovetter. The Strength of Weak Ties. *American Journal of Sociology*, 78(6):1360, 1973. DOI: 10.1086/225469 18

M. Granovetter. Threshold models of collective behavior. *American journal of sociology*, 83(6):1420, 1978. DOI: 10.1086/226707 22

D. Gruhl, R. Guha, D. Liben-Nowell, and A. Tomkins. Information diffusion through blogspace. In *WWW '04: Proceedings of the 13th international conference on World Wide Web*, pages 491–501, New York, NY, USA, 2004. ACM. ISBN 1-58113-844-X. DOI: 10.1145/988672.988739 23

M. S. Handcock, A. E. Raftery, and J. M. Tantrum. Model-based clustering for social networks. *Journal of The Royal Statistical Society Series A*, 127(2):301–354, 2007. URL http://ideas.repec.org/a/bla/jorssa/v170y2007i2p301-354.html. 37

R. Hanneman and M. Riddle. *Introduction to Social Network Methods*. http://faculty.ucr.edu/ hanneman/, 2005. 36

M. Hechter. *Principles of Group Solidarity*. University of California Press, 1988. 8

S. Hill, F. Provost, and C. Volinsky. Network-based marketing: Identifying likely adopters via consumer networks. *Statistical Science*, 21(2):256–276, 2006. DOI: 10.1214/088342306000000222 29

P. D. Hoff, A. E. Raftery, and M. S. Handcock. Latent space approaches to social network analysis. *Journal of the American Statistical Association*, 97(460):1090–1098, 2002. DOI: 10.1198/016214502388618906 37

J. Hopcroft and R. Tarjan. Algorithm 447: efficient algorithms for graph manipulation. *Commun. ACM*, 16(6):372–378, 1973. ISSN 0001-0782. DOI: 10.1145/362248.362272 33

J. Hopcroft, O. Khan, B. Kulis, and B. Selman. Natural communities in large linked networks. In *KDD '03: Proceedings of the ninth ACM SIGKDD international conference on Knowledge discovery and data mining*, pages 541–546, New York, NY, USA, 2003. ACM. ISBN 1-58113-737-0. DOI: 10.1145/956750.956816 36, 78

J. Hopcroft, O. Khan, B. Kulis, and B. Selman. Tracking evolving communities in large linked networks. *Proceedings of the National Academy of Sciences of the United States of America*, 101 (Suppl 1):5249–5253, 2004. URL http://www.pnas.org/content/101/suppl.1/5249. abstract. DOI: 10.1073/pnas.0307750100 78, 79

B. A. Huberman, D. M. Romero, and F. Wu. Social networks that matter: Twitter under the microscope. *First Monday*, 14(1), 2009. 20

A. Java, A. Joshi, and T. Finin. Detecting Commmunities via Simultaneous Clustering of Graphs and Folksonomies. In *Proceedings of the Tenth Workshop on Web Mining and Web Usage Analysis (WebKDD)*. ACM, August 2008. (Held in conjunction with The 14th ACM SIGKDD International Conference on Knowledge Discovery and Data Mining (KDD 2008)). 9

D. B. Johnson. Efficient algorithms for shortest paths in sparse networks. *J. ACM*, 24(1):1–13, 1977. ISSN 0004-5411. DOI: 10.1145/321992.321993 18

D. Kempe, J. Kleinberg, and É. Tardos. Maximizing the spread of influence through a social network. In *Proceedings of the ninth ACM SIGKDD international conference on Knowledge discovery and data mining*, pages 137–146. ACM New York, NY, USA, 2003. DOI: 10.1145/956750.956769 8, 21, 23, 24, 25, 26

M.-S. Kim and J. Han. A particle-and-density based evolutionary clustering method for dynamic networks. *Proc. VLDB Endow.*, 2(1):622–633, 2009. ISSN 2150-8097. 82

P. Kolari, T. Finin, and A. Joshi. SVMs for the blogosphere: Blog identification and splog detection. In *AAAI Spring Symposium on Computational Approaches to Analyzing Weblogs*, 2006a. 11

P. Kolari, A. Java, and T. Finin. Characterizing the splogosphere. In *Proceedings of the 3rd Annual Workshop on Weblogging Ecosystem: Aggregation, Analysis and Dynamics, 15th World Wid Web Conference*. Citeseer, 2006b. 11

G. Kossinets and D. J. Watts. Empirical analysis of an evolving social network. *Science*, 311(5757): 88–90, 2006. URL http://www.sciencemag.org/cgi/content/abstract/311/5757/88. DOI: 10.1126/science.1116869 76

G. Kossinets, J. Kleinberg, and D. Watts. The structure of information pathways in a social communication network. In *KDD '08: Proceeding of the 14th ACM SIGKDD international conference on Knowledge discovery and data mining*, pages 435–443, New York, NY, USA, 2008. ACM. ISBN 978-1-60558-193-4. DOI: 10.1145/1401890.1401945 21

R. Kumar, P. Raghavan, S. Rajagopalan, and A. Tomkins. Trawling the web for emerging cyber-communities. *Comput. Netw.*, 31(11-16):1481–1493, 1999. ISSN 1389-1286. DOI: 10.1016/S1389-1286(99)00040-7 32

R. Kumar, J. Novak, P. Raghavan, and A. Tomkins. On the bursty evolution of blogspace. *World Wide Web*, 8(2):159–178, 2005. ISSN 1386-145X. DOI: 10.1007/s11280-004-4872-4 75, 76, 78

R. Kumar, J. Novak, and A. Tomkins. Structure and evolution of online social networks. In *KDD '06: Proceedings of the 12th ACM SIGKDD international conference on Knowledge discovery and data mining*, pages 611–617, New York, NY, USA, 2006. ACM. ISBN 1-59593-339-5. DOI: 10.1145/1150402.1150476 33, 76

T. La Fond and J. Neville. Randomization tests for distinguishing social influence and homophily effects. In *WWW '10: Proceedings of the 19th international conference on World wide web*, pages 601–610, New York, NY, USA, 2010. ACM. ISBN 978-1-60558-799-8. DOI: 10.1145/1772690.1772752 27, 29

J. Leskovec and C. Faloutsos. Sampling from large graphs. In *KDD '06: Proceedings of the 12th ACM SIGKDD international conference on Knowledge discovery and data mining*, pages 631–636, New York, NY, USA, 2006. ACM. ISBN 1-59593-339-5. DOI: 10.1145/1150402.1150479 8

J. Leskovec and E. Horvitz. Planetary-scale views on a large instant-messaging network. In *WWW '08: Proceeding of the 17th international conference on World Wide Web*, pages 915–924, New York, NY, USA, 2008. ACM. ISBN 978-1-60558-085-2. DOI: 10.1145/1367497.1367620 5, 6

J. Leskovec, J. Kleinberg, and C. Faloutsos. Graph evolution: Densification and shrinking diameters. *ACM Trans. Knowl. Discov. Data*, 1(1):2, 2007a. ISSN 1556-4681. DOI: 10.1145/1217299.1217301 76

J. Leskovec, A. Krause, C. Guestrin, C. Faloutsos, J. VanBriesen, and N. Glance. Cost-effective outbreak detection in networks. In *KDD '07: Proceedings of the 13th ACM SIGKDD international conference on Knowledge discovery and data mining*, pages 420–429, New York, NY, USA, 2007b. ACM. ISBN 978-1-59593-609-7. DOI: 10.1145/1281192.1281239 8, 24, 26

J. Leskovec, L. Backstrom, R. Kumar, and A. Tomkins. Microscopic evolution of social networks. In *KDD '08: Proceeding of the 14th ACM SIGKDD international conference on Knowledge discovery and data mining*, pages 462–470, New York, NY, USA, 2008. ACM. ISBN 978-1-60558-193-4. DOI: 10.1145/1401890.1401948 76

D. Liben-Nowell and J. Kleinberg. The link-prediction problem for social networks. *J. Am. Soc. Inf. Sci. Technol.*, 58(7):1019–1031, 2007. ISSN 1532-2882. DOI: 10.1002/asi.20591 10, 11

Y.-R. Lin, Y. Chi, S. Zhu, H. Sundaram, and B. L. Tseng. Analyzing communities and their evolutions in dynamic social networks. *ACM Trans. Knowl. Discov. Data*, 3(2):1–31, 2009. ISSN 1556-4681. DOI: 10.1145/1514888.1514891 82

B. Long, Z. M. Zhang, X. Wú, and P. S. Yu. Spectral clustering for multi-type relational data. In *ICML '06: Proceedings of the 23rd international conference on Machine learning*, pages 585–592, New York, NY, USA, 2006. ACM. ISBN 1-59593-383-2. DOI: 10.1145/1143844.1143918 72

B. Long, P. S. Yu, and Z. M. Zhang. A general model for multiple view unsupervised learning. In *SDM '08: Proceedings of SIAM International Conference on Data Mining*, pages 822–833, 2008. 64

U. v. Luxburg. A tutorial on spectral clustering. *Statistics and Computing*, 17(4):395–416, 2007. ISSN 0960-3174. DOI: 10.1007/s11222-007-9033-z 41, 42

S. A. Macskassy and F. Provost. Classification in networked data: A toolkit and a univariate case study. *J. Mach. Learn. Res.*, 8:935–983, 2007. ISSN 1533-7928. 85, 86

S. C. Madeira and A. L. Oliveira. Biclustering algorithms for biological data analysis: A survey. *IEEE/ACM Transactions on Computational Biology and Bioinformatics*, 1:24–45, 2004. ISSN 1545-5963. DOI: 10.1109/TCBB.2004.2 69

B. McClosky and I. V. Hicks. Detecting cohesive groups. http://www.caam.rice.edu/ ivhicks/CokplexAlgorithmPaper.pdf, 2009. 34

M. McPherson, L. Smith-Lovin, and J. M. Cook. Birds of a feather: Homophily in social networks. *Annual Review of Sociology*, 27:415–444, 2001. DOI: 10.1146/annurev.soc.27.1.415 27

F. Menczer. Web crawling. In B. Liu, editor, *Web Data Mining*, chapter 8, pages 273–322. Springer, 2006. 94

A. Mislove, M. Marcon, K. P. Gummadi, P. Druschel, and B. Bhattacharjee. Measurement and analysis of online social networks. In *IMC '07: Proceedings of the 7th ACM SIGCOMM conference on Internet measurement*, pages 29–42, New York, NY, USA, 2007. ACM. ISBN 978-1-59593-908-1. DOI: 10.1145/1298306.1298311 6

T. M. Mitchell. Mining Our Reality. *Science*, 326(5960):1644–1645, 2009. URL http://www.sciencemag.org. DOI: 10.1126/science.1174459 11

A. A. Nanavati, S. Gurumurthy, G. Das, D. Chakraborty, K. Dasgupta, S. Mukherjea, and A. Joshi. On the structural properties of massive telecom call graphs: findings and implications. In *CIKM '06: Proceedings of the 15th ACM international conference on Information and knowledge management*, pages 435–444, New York, NY, USA, 2006. ACM. ISBN 1-59593-433-2. DOI: 10.1145/1183614.1183678 6

G. Nemhauser, L. Wolsey, and M. Fisher. An analysis of approximations for maximizing submodular set functions-I. *Mathematical Programming*, 14(1):265–294, 1978. DOI: 10.1007/BF01588971 26

M. Newman. Modularity and community structure in networks. *PNAS*, 103(23):8577–8582, 2006a. DOI: 10.1073/pnas.0601602103 43, 53

M. Newman. Finding community structure in networks using the eigenvectors of matrices. *Physical Review E (Statistical, Nonlinear, and Soft Matter Physics)*, 74(3), 2006b. URL http://dx.doi.org/10.1103/PhysRevE.74.036104. DOI: 10.1103/PhysRevE.74.036104 44, 46, 49

M. Newman and M. Girvan. Finding and evaluating community structure in networks. *Physical Review E*, 69:026113, 2004. URL http://www.citebase.org/abstract?id=oai:arXiv.org:cond-mat/0308217. DOI: 10.1103/PhysRevE.69.026113 9, 46, 97

M. Newman, A.-L. Barabasi, and D. J. Watts, editors. *The Structure and Dynamics of Networks*. 2006. 6

J. Onnela, J. Saramäki, J. Hyvönen, G. Szabó, D. Lazer, K. Kaski, J. Kertész, and A. Barabási. Structure and tie strengths in mobile communication networks. *PNAS*, 104(18):7332–7336, 2007. DOI: 10.1073/pnas.0610245104 19

L. Page, S. Brin, R. Motwani, and T. Winograd. The PageRank citation ranking: Bringing order to the web. Technical Report 1999-66, Stanford InfoLab, November 1999. URL http://ilpubs.stanford.edu:8090/422/. Previous number = SIDL-WP-1999-0120. 8, 16

G. Palla, I. Derényi, I. Farkas, and T. Vicsek. Uncovering the overlapping community structure of complex networks in nature and society. *Nature*, 435:814–818, 2005. DOI: 10.1038/nature03607 33, 78

G. Palla, A.-L. Barabasi, and T. Vicsek. Quantifying social group evolution. *Nature*, 446(7136): 664–667, April 2007. DOI: 10.1038/nature05670 9, 76, 77, 78

J. Pearl. *Causality: models, reasoning, and inference*. Cambridge University Press, 2000. 28

M. Ramezani, J. Sandvig, R. Bhaumik, R. Burke, and B. Mobasher. Exploring the impact of profile injection attacks in social tagging systems. In *Proceedings of Workshop on Web Mining and Web Usage Analysis*, 2008. 11

M. Richardson and P. Domingos. Mining knowledge-sharing sites for viral marketing. In *KDD '02: Proceedings of the eighth ACM SIGKDD international conference on Knowledge discovery and data mining*, pages 61–70, New York, NY, USA, 2002. ACM. ISBN 1-58113-567-X. DOI: 10.1145/775047.775057 8, 24

K. Saito, M. Kimura, K. Ohara, and H. Motoda. Behavioral analyses of information diffusion models by observed data of social network. In *SBP*, pages 149–158, 2010. DOI: 10.1007/978-3-642-12079-4_20 21

P. Sarkar and A. W. Moore. Dynamic social network analysis using latent space models. *SIGKDD Explor. Newsl.*, 7(2):31–40, 2005. ISSN 1931-0145. DOI: 10.1145/1117454.1117459 37, 82

T. C. Schelling. Dynamic models of segregation. *Journal of Mathematical Sociology*, 1:143–186, 1971. DOI: 10.1080/0022250X.1971.9989794 22

P. Sen, G. Namata, M. Bilgic, L. Getoor, B. Galligher, and T. Eliassi-Rad. Collective classification in network data. *AI Magazine*, 29(3):93, 2008. 85

C. Shirky. *Here Comes Everybody: The Power of Organizing without Organizations*. The Penguin Press, 2008. 1

P. Singla and M. Richardson. Yes, there is a correlation: - from social networks to personal behavior on the web. In *WWW '08: Proceeding of the 17th international conference on World Wide Web*, pages 655–664, New York, NY, USA, 2008. ACM. ISBN 978-1-60558-085-2. DOI: 10.1145/1367497.1367586 26

A. Strehl and J. Ghosh. Cluster ensembles — a knowledge reuse framework for combining multiple partitions. *J. Mach. Learn. Res.*, 3:583–617, 2003. ISSN 1533-7928. DOI: 10.1162/153244303321897735 49, 65

J. Sun, C. Faloutsos, S. Papadimitriou, and P. S. Yu. Graphscope: parameter-free mining of large time-evolving graphs. In *KDD '07: Proceedings of the 13th ACM SIGKDD international conference on Knowledge discovery and data mining*, pages 687–696, New York, NY, USA, 2007. ACM. ISBN 978-1-59593-609-7. DOI: 10.1145/1281192.1281266 82, 83

Y. Sun, Y. Yu, and J. Han. Ranking-based clustering of heterogeneous information networks with star network schema. In *KDD '09: Proceedings of the 15th ACM SIGKDD international conference on Knowledge discovery and data mining*, pages 797–806, New York, NY, USA, 2009. ACM. ISBN 978-1-60558-495-9. DOI: 10.1145/1557019.1557107 56

P.-N. Tan, M. Steinbach, and V. Kumar. *Introduction to Data Mining*. Addison Wesley, 2005. 36, 37, 49, 85, 101

L. Tang. *Learning with Large-Scale Social Media Networks*. PhD thesis, Arizona State University, 2010. URL http://www.public.asu.edu/~ltang9/thesis.pdf. 52, 58

L. Tang and H. Liu. Scalable learning of collective behavior based on sparse social dimensions. In *CIKM '09: Proceeding of the 18th ACM conference on Information and knowledge management*, pages 1107–1116, New York, NY, USA, 2009a. ACM. ISBN 978-1-60558-512-3. DOI: 10.1145/1645953.1646094 91

L. Tang and H. Liu. Relational learning via latent social dimensions. In *KDD '09: Proceedings of the 15th ACM SIGKDD international conference on Knowledge discovery and data mining*, pages 817–826, New York, NY, USA, 2009b. ACM. ISBN 978-1-60558-495-9. DOI: 10.1145/1557019.1557109 57, 88

L. Tang and H. Liu. Toward predicting collective behavior via social dimension extraction. *IEEE Intelligent Systems*, 25:19–25, 2010a. ISSN 1541-1672. DOI: 10.1109/MIS.2010.36 89, 90

L. Tang and H. Liu. Graph mining applications to social network analysis. In C. Aggarwal and H. Wang, editors, *Managing and Mining Graph Data*, chapter 16, pages 487–513. Springer, 2010b. DOI: 10.1007/978-1-4419-6045-0_16 31

L. Tang, H. Liu, J. Zhang, and Z. Nazeri. Community evolution in dynamic multi-mode networks. In *KDD '08: Proceeding of the 14th ACM SIGKDD international conference on Knowledge discovery and data mining*, pages 677–685, New York, NY, USA, 2008. ACM. ISBN 978-1-60558-193-4. DOI: 10.1145/1401890.1401972 9, 49, 72

L. Tang, X. Wang, and H. Liu. Uncovering groups via heterogeneous interaction analysis. In *ICDM '09: Proceedings of IEEE International Conference on Data Mining*, pages 503–512, 2009. 9, 46, 55, 64, 67

L. Tang, X. Wang, H. Liu, and L. Wang. A multi-resolution approach to learning with overlapping communities. In *Proceedings of Workshop on Social Media Analytics*, 2010. 92

L. Tang, H. Liu, and J. Zhang. Identifying evolving groups in dynamic multi-mode networks. *IEEE Transactions on Knowledge and Data Engineering (TKDE)*, forthcoming. 9, 56, 73

C. Tantipathananandh, T. Berger-Wolf, and D. Kempe. A framework for community identification in dynamic social networks. In *KDD '07: Proceedings of the 13th ACM SIGKDD international conference on Knowledge discovery and data mining*, pages 717–726, New York, NY, USA, 2007. ACM. ISBN 978-1-59593-609-7. DOI: 10.1145/1281192.1281269 82

M. Thelwall. Bloggers under the london attacks:top information sources and topics. In *WWW:3rd annual workshop on webloging ecosystem: aggreation, analysis and dynamics*, 2006. 2

J. Travers and S. Milgram. An experimental study of the small world problem. *Sociometry*, 32(4): 425–443, 1969. DOI: 10.2307/2786545 5

K. Wakita and T. Tsurumi. Finding community structure in mega-scale social networks: [extended abstract]. In *WWW '07: Proceedings of the 16th international conference on World Wide Web*, pages 1275–1276, New York, NY, USA, 2007. ACM. ISBN 978-1-59593-654-7. DOI: 10.1145/1242572.1242805 48

S. Wasserman and K. Faust. *Social Network Analysis: Methods and Applications*. Cambridge University Press, 1994. 3, 5, 8, 13, 34, 68

D. J. Watts and P. S. Dodds. Influentials, networks, and public opinion formation. *Journal of Consumer Research*, 34(4):441–458, 2007. DOI: 10.1086/518527 26

D. J. Watts and S. H. Strogatz. Collective dynamics of 'small-world' networks. *Nature*, 393:440–442, 1998. DOI: 10.1038/30918 7, 8

R. Xiang, J. Neville, and M. Rogati. Modeling relationship strength in online social networks. In *WWW '10: Proceedings of the 19th international conference on World wide web*, pages 981–990, New York, NY, USA, 2010. ACM. ISBN 978-1-60558-799-8. DOI: 10.1145/1772690.1772790 20

T. Yang, Y. Chi, S. Zhu, Y. Gao, and R. Jin. A bayesian approach toward finding communities and their evolutions in dynamic social networks. In *SDM '09: Proceedings of SIAM International Conference on Data Mining*, 2009. 82

H.-J. Zeng, Z. Chen, and W.-Y. Ma. A unified framework for clustering heterogeneous web objects. In *WISE '02: Proceedings of the 3rd International Conference on Web Information Systems Engineering*, pages 161–172, Washington, DC, USA, 2002. IEEE Computer Society. ISBN 0-7695-1766-8. DOI: 10.1109/WISE.2002.1181653 9

X. Zhu, Z. Ghahramani, and J. Lafferty. Semi-supervised learning using gaussian fields and harmonic functions. In *ICML '03: Proceedings of the Twentieth International Conference on Machine Learning*, pages 912–919, 2003. 87

Authors' Biographies

LEI TANG

Lei Tang is a scientist at Yahoo! Labs[1]. He received his Ph.D. in computer science and engineering at Arizona State University in 2010 and BS from Fudan University, China in 2004. His research interests include social computing, data mining, and social media mining, in particular, relational learning with heterogeneous networks, group evolution, profiling and influence modeling, and collective behavior modeling and prediction in social media. He was awarded ASU GPSA Research Grant, SDM Doctoral Student Forum Fellowship, Student Travel Awards and Scholarships in various conferences. He is a member of ACM and IEEE.

HUAN LIU

Huan Liu is a professor of computer science and engineering at Arizona State University (ASU). He received his Ph.D. from University of Southern California and Bachelor of Engineering from Shanghai Jiao Tong University. He has been recognized for excellence in teaching and research in the Departement of Computer Science and Engineering at ASU. His research interests include data/web mining, machine learning, social computing, and artificial intelligence, investigating problems that arise in many real-world applications with high-dimensional data of disparate forms and multiple sources such as feature selection, modeling group interaction, relational learning, text categorization, biomarker identification, and social media analysis. His well-cited publications include books, book chapters, encyclopedia entries, conference and journal papers. He serves on journal editorial boards and numerous conference program committees, and he is a founding organizer of the International Workshop/Conference Series on Social Computing, Behavioral Modeling, and Prediction (SBP), http://sbp.asu.edu/. His professional memberships include AAAI, ACM, ASEE, and IEEE. He can be contacted via http://www.public.asu.edu/~huanliu.

---

[1]This book was completed when Lei was at Arizona State University.

Index

Printed in the United States
by Baker & Taylor Publisher Services